AS: USE OF MATHS

Statistics

JUNE HAIGHTON • ANNE HAWORTH • GEOFF WAKE

Cost per week (£)	Number of students
0–5	113
5–10	202
10–15	215
15–20	215
More than 20	127
Total	**872**

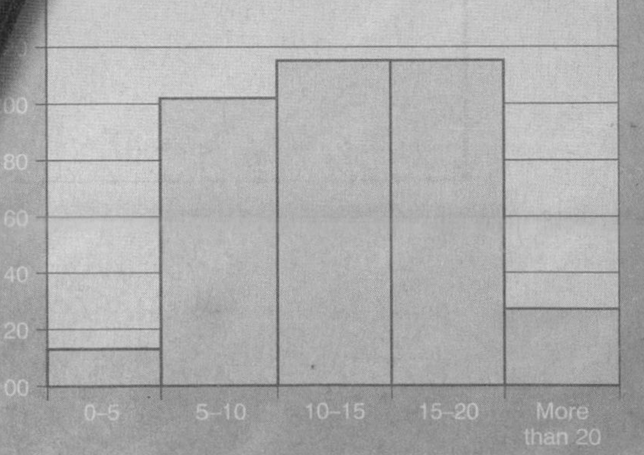

Nelson Thornes
a Wolters Kluwer business

Published in 2003 by:
Nelson Thornes Ltd
Delta Place
27 Bath Road
CHELTENHAM
GL53 7TH
United Kingdom

08 09 10 / 10 9 8 7 6 5

A catalogue record for this book is available from the British Library

ISBN 978 0 7487 6980 3

Illustrations by Angela Lumley
Page make-up by Tech Set Ltd

Printed and bound in Slovenia by Korotan-Ljubljana

Acknowledgements

The development of this publication was supported by the Nuffield Foundation
www.fsmq.org

The authors would like to thank Lawrence Wo for his help and assistance in checking and compiling the manuscript

The publishers wish to thank the following for permission to reproduce copyright material: McDonald's Archives; Foxx (NT): vi; Stockbyte 9 (NT): 3; Photodisc 22 (NT): 7; Corel 747 (NT): 12; Digital Stock 1 (NT): 15; Corel 772 (NT): 37; Digital Stock 10 (NT): 64; Photodisc 59 (NT): 84; Corel 790 (NT): 129; Activity Games: 2. All other photographs Nelson Thornes Archive.

The publishers have made every effort to contact copyright holders but apologise if any have been overlooked.

Contents

Using this book

Statistics is a branch of mathematics that allows you to describe and make sense of numerical data gathered from the world about you.

You will already know how to calculate some statistical measures, such as averages, and how to draw statistical diagrams, such as bar charts and pie charts.

However, statistics is about much more than that. You should use this book to assist you in carrying out investigations in which you:

- identify an area or problem to investigate
- select appropriate data to use
- carry out data analysis
- draw conclusions and summarise findings.

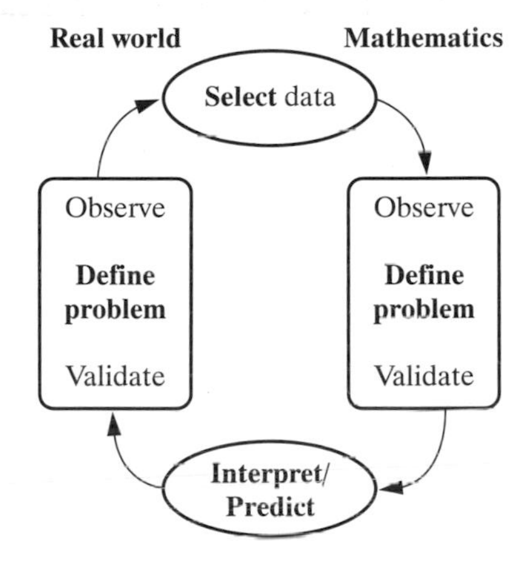

This diagram attempts to summarise this process.

You will need to be involved in all aspects of this process when completing work for your coursework portfolio.

Throughout your work you need to have access to technology. It is often useful to work with large sets of data using a spreadsheet. Some of the activities in this book suggest that you use a spreadsheet – at times your teacher may have a file with the data entered ready for you to use.

You will also need to use a calculator that can calculate statistical measures such as correlation coefficients. Many scientific calculators will this – you may find a graphic calculator useful as these also allow you to draw statistical diagrams as well as calculating all the measures you need.

See the section 'Using Technology' to get some advice.

Anthropometric data for 14–15 year olds – see Chapter 2.

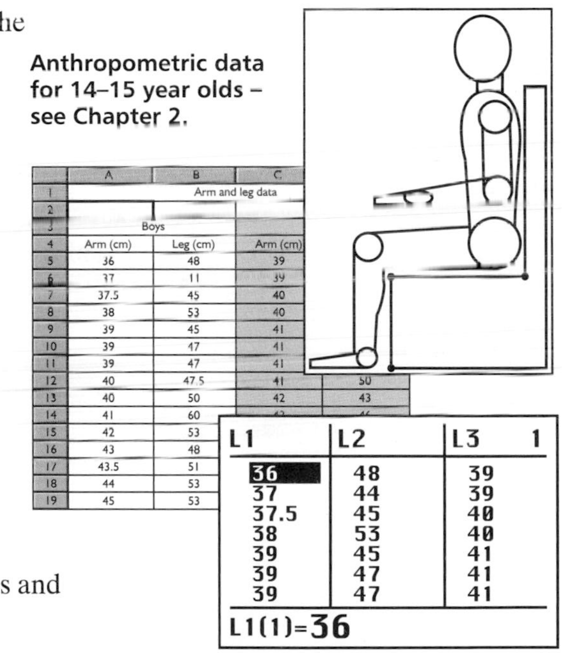

Using a graphic calculator to calculate statistical measures and draw diagrams

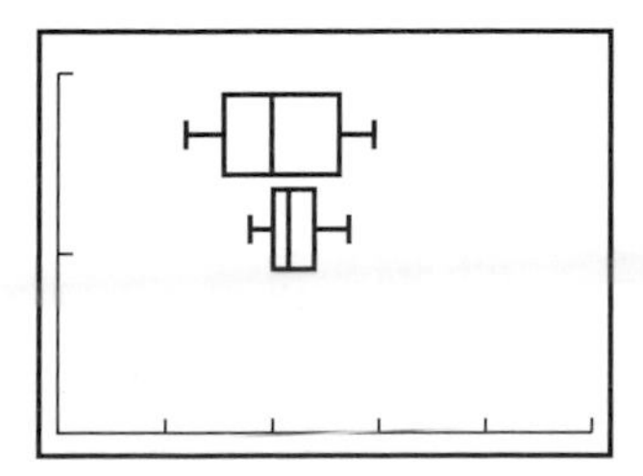

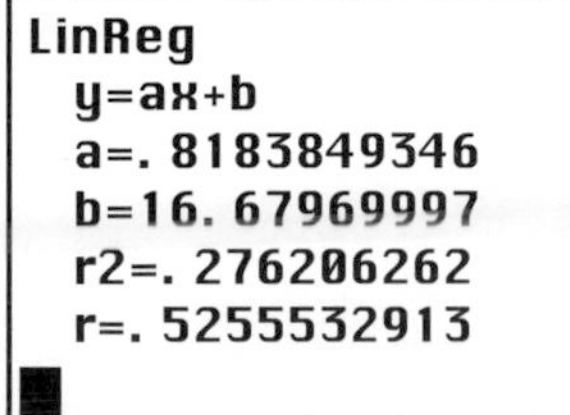

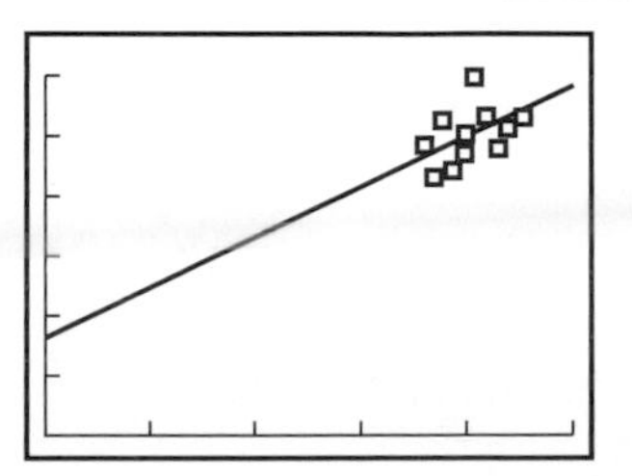

Introduction

Statistics has a bad reputation. In the past many organisations, such as tobacco, cosmetics and drugs companies as well as all sorts of pressure groups have used statistics in a misleading way to support their claims. Others have been accused of 'being economical with the truth', giving partial information that excludes any data that contradicts their claims. Even the government has been said to put a 'spin' on official figures. They have been accused of sometimes double-counting data or omitting to mention those that show their policies in a bad light.

A much-quoted phrase comes at the end of this passage from Mark Twain's autobiography:

> *'Figures often beguile me, particularly when I have the arranging of them myself; in which case the remark attributed to Disraeli would often apply with justice and force: "There are three kinds of lies: lies, damned lies and statistics." '*

The point being made here of course is that statistics can be manipulated to misrepresent facts to support a particular agenda. However, statistical methods arm us with an array of extremely valuable tools when used correctly. For example, medical researchers use statistical methods to tell them whether a new drug is having a beneficial effect or not, manufacturers use statistics to control the quality of the goods they produce, investors use statistics to make decisions about where to put their money and the person in the street uses statistics to see how well his or her football team is doing.

In this book you will learn how to find statistical measures and how to represent data using a range of diagrams. You will learn how to explore the relationships between variables and how to model data using the Normal distribution. You will also consider critically the statistics presented by others.

What the scientists said:
There is a *'real association between carcinoma of the lung and smoking'*.
(Dr Richard Doll and Prof Barbara Hill 1952)

What the tobacco companies said:
'There is no proof that cigarette smoking is one of the causes' of lung cancer.
(Tobacco Industry Research Committee 1954)

Government 'Spin'

Both Labour and Conservative governments have been accused of manipulating statistics.

The 1980s' Thatcher government restricted access to social benefits. This meant that official unemployment figures, based on the monthly claimant count, decreased and gave a false impression.

More recently, the Blair government was accused of double-counting the money it had allocated to the National Health Service.

The government's Rough Sleepers Unit was also accused of fixing the count of homeless people by moving rough sleepers into hostels for a single night so that it could meet its target.

Drug trial on diabetic heart patients

What the researchers said:
About 5 years of treatment can prevent heart attacks, strokes and other major vascular events.
Estimate of the number of lives potentially saved in the UK per year = 5000.
(University of Oxford)

What the papers said:
Estimates of lives potentially saved = 25 000.
(*Daily Telegraph and Daily Mail*)

This brief introductory chapter starts with a fictitious report using statistical ideas which you may have already met. Read it carefully. Discuss with others all the faults.

Mobile phone ban

Message from the Principal **Mobile Phone Ban**

Today I have taken the decision to ban the use of mobile phones by students in this college. The decision has been taken after a survey to find the extent to which the use of mobile phones is affecting the education of students here. I have also studied recent research and advice from the Department of Health about the use of mobile phones. The results of the college survey and relevant extracts from Government advice are available in a newsletter. I hope that you will take time to read this and agree with my decision to ban mobile phones in the light of the evidence given.

Newsletter Special

College Mobile Phone Survey

Introduction
Over the past year the use of mobile phones in college has increased enormously. There have been complaints from staff about mobile phones disrupting both classes and examinations. Concerns about the effects of mobile phones on health have been expressed both nationally and in college. Recently a survey has been carried out in college to find out the extent to which students use mobile phones and to try to ascertain the effects that this has on their education.

Methods
Lecturers were asked to distribute copies of the following short questionnaire to any student who they saw using a mobile phone before, during or after lectures. Other staff were employed to patrol college grounds during breaks distributing questionnaires to any student seen using a mobile phone. In this way data was collected from a wide variety of students. The total number of students who completed questionnaires was 872.

Discussion points
Do you think that this was a good way to distribute the questionnaires? Can you suggest better methods?

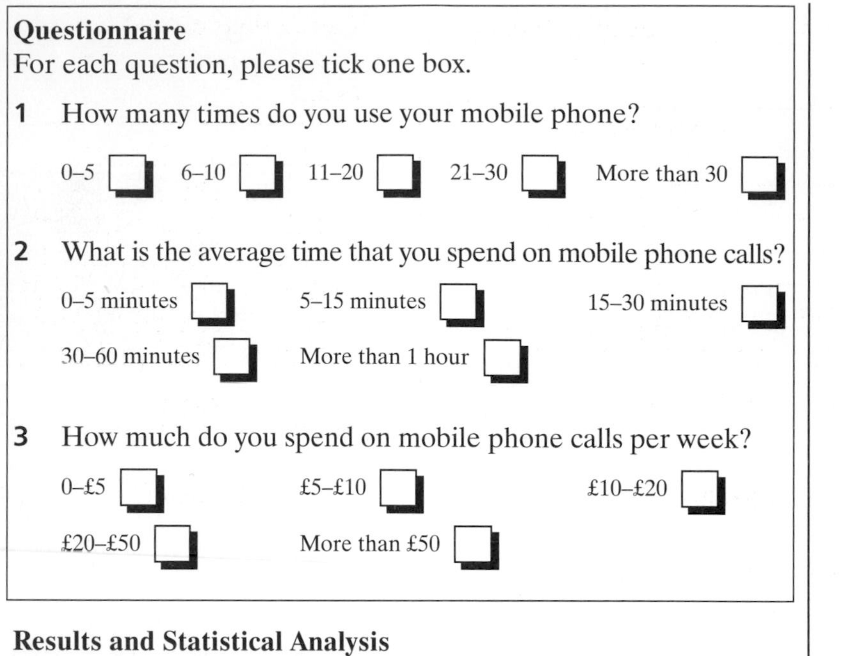

Questionnaire

For each question, please tick one box.

1 How many times do you use your mobile phone?

0–5 ☐ 6–10 ☐ 11–20 ☐ 21–30 ☐ More than 30 ☐

2 What is the average time that you spend on mobile phone calls?

0–5 minutes ☐ 5–15 minutes ☐ 15–30 minutes ☐

30–60 minutes ☐ More than 1 hour ☐

3 How much do you spend on mobile phone calls per week?

0–£5 ☐ £5–£10 ☐ £10–£20 ☐

£20–£50 ☐ More than £50 ☐

Discussion points

Are these questions

- easy to answer
- unambiguous
- likely to get a truthful response?

Results and Statistical Analysis

The results from each question on the questionnaire are given in the tables and charts below.

Question 1 How many times do you use your mobile phone

Number of calls	Number of students
0–5	51
6–10	328
11–20	385
21–30	96
More than 30	12
Total	**872**

The mean number of calls per student was calculated to be 13.7.

Calculations carried out to find the mean are given below:

x	f	xf
2.5	51	127.5
8.0	328	2624.0
15.5	385	5967.5
25.5	96	2448.0
65.0	12	780.0
Total	**872**	**11 947.0**

$$\text{Mean} = \frac{11\,947}{872} = 13.7 \text{ (1 d.p.)}$$

Discussion points

Could you explain how the calculations above have been carried out to find the mean?

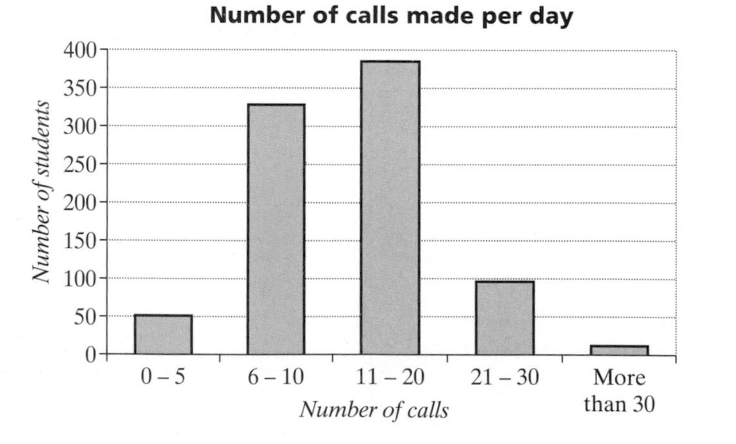

The majority of students attending this college make between 11 and 20 mobile phone calls per day.

Discussion points

Is the mean a good representative of this data?

Does the bar chart illustrate the data fairly?

Is it true that the majority of students attending the college make between 11 and 20 mobile phone calls per day?

Question 2 What is the average time that you spend on mobile phone calls?

Length of calls	Number of calls
0–5	134
5–15	317
15–30	323
30–60	96
More than 60	2
Total	**872**

The mean length of all calls made by college students was calculated to be 19.0 minutes.

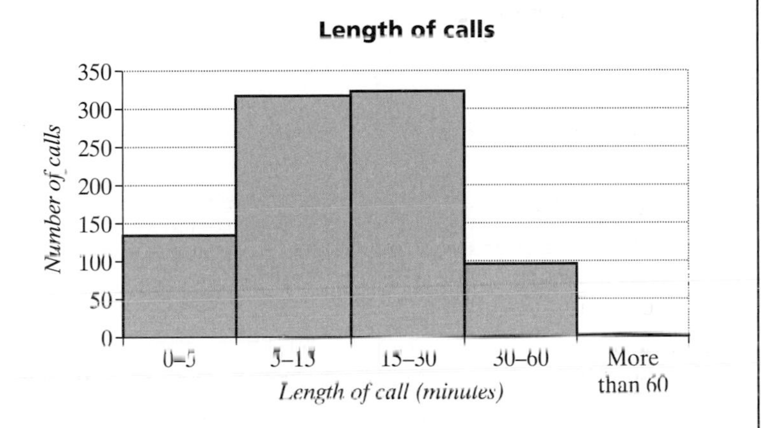

The majority of mobile phone calls made by college students last between 15 and 30 minutes.

Question 3 How much do you spend on mobile phone calls per week?

Cost per week (£)	Number of students
0–5	113
5–10	202
10–15	215
15–20	215
More than 20	127
Total	**872**

The mean cost per week was calculated to be £18.20.

Calculations carried out to find the mean are given below:

x	*f*	*xf*
2.5	134	335.0
10.0	317	3170.0
22.5	323	7267.5
45.0	96	4320.0
750.0	2	1500.0
Total	**872**	**16 592.5**

$$\text{Mean} = \frac{16\,592.5}{872}$$

$$= 19.0 \text{ min (1 d.p.)}$$

Discussion points

Can you explain the calculation above?

Is the mean length of all calls made by college students really 19.0 minutes?

Does the diagram illustrate the data fairly?

Is it true that the majority of mobile phone calls made by college students last between 15 and 30 minutes?

Calculations carried out to find the mean are given below:

x	*f*	*xf*
2.5	113	282.5
7.5	202	1515.0
12.5	215	2687.5
17.5	215	3762.5
60.0	127	7620.0
Total	**872**	**15 867.5**

$$\text{Mean} = \frac{15\,867.5}{872}$$

$$= £18.20 \text{ (2 d.p.)}$$

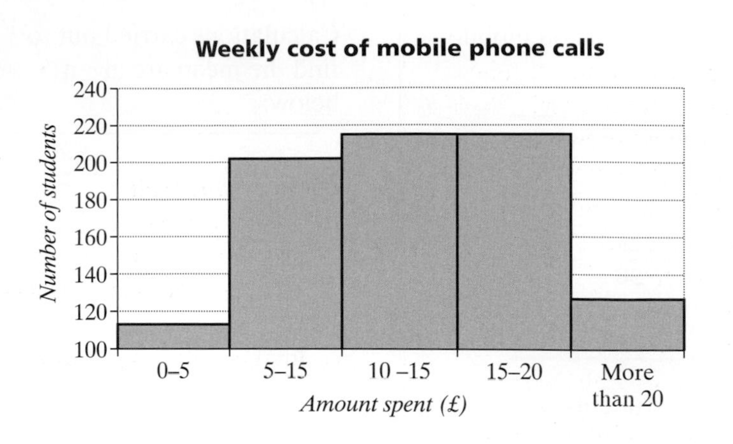

Most students spend between £10 and £20 on mobile phone calls per week.

Class Interruptions

In addition, all lecturers have been asked whether or not a mobile phone has ever interrupted one of their classes. The result (73 out of 132 lecturers reported interruptions) is illustrated by the pie chart.

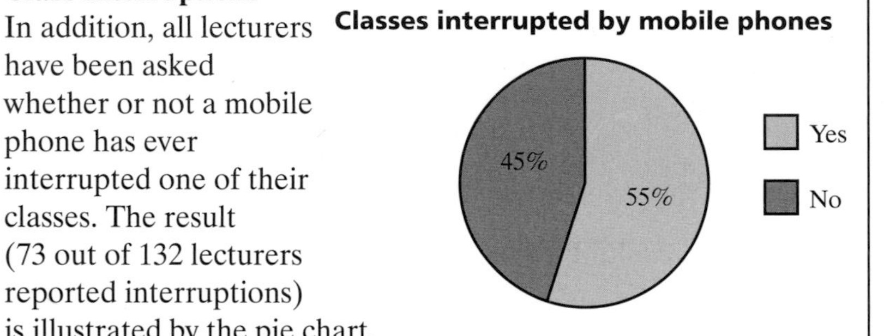

Over half of classes have been interrupted by a mobile phone.

Summary of Findings from College Mobile Phone Survey

- The majority of students attending this college make between 11 and 20 mobile phone calls per day, with the mean number of mobile phone calls per student per day being 13.7.
- The majority of mobile phone calls made by college students last between 15 and 30 minutes, the mean length being 19.0 minutes.
- Most students spend between £10 and £20 on mobile phone calls per week, with a mean cost per week of £18.20.
- Over half of classes are interrupted by mobile phone calls.

Advice from the Department of Health

Commenting on recent research the Department of Health states:

'*The research does show that using mobile phones affects brain activity. The UK Chief Medical Officers strongly advise that where young people do use mobile phones, they should be encouraged to:*

- *use mobile phones for essential purposes only*
- *keep all calls short – talking for long periods prolongs exposure to radio frequency radiation and should be discouraged.*'

Discussion points

Can you explain the calculations used to find the mean cost per week?

Is the mean a good representative of this data?

Does the diagram illustrate the data fairly?

Is it true that most students spend between £10 and £20 on mobile phone calls per week?

Discussion points

Do you think the pie chart gives a true reflection of problems caused by mobile phones interrupting classes?

Is it true that over half of classes have been interrupted by a mobile phone?

Visit the Department of Health's website to see all the information and advice given about mobile phones.

Discussion point

Do you think that this is a fair summary of the advice given by the Department of Health?

Principal's Comments on Findings
I am alarmed at the extent to which students are using mobile phones in college. The college survey shows that the majority of students are making a large number of calls and spending long periods of time on mobile phones each day. The costs incurred through the use of mobile phones are substantial and many students are likely to have increased significantly the time they spend doing part-time work to finance this cost. This, in addition to the time spent using mobile phones, will cause a reduction in the time available for studies. There is also widespread disruption to classes because of the use of mobile phones.

Information from the Department of Health suggests that there are health risks associated with the use of mobile phones. It would be extremely difficult to enforce the Department of Health's recommendations for the safe use of mobile phones in college.

In view of these facts I have decided that a complete ban on mobile phones is the only sensible way to ensure that the health and education of the students at this college do not suffer because of the use of mobile phones.

From the start of next term:

STUDENTS MUST NOT BRING MOBILE PHONES INTO THIS COLLEGE

A pay phone will be installed in each of the college buildings to enable students to make calls during breaks.

Discussion point
Which, if any, of the principal's comments do you agree with?
Which comments do you disagree with?
Which comments are justified?

1 Exploring and Describing Data

Contents

This article at the BBC News website explains how even the 2001 UK census did not in fact have information about everyone in the country. The Office of National Statistics has complete census forms from only a **sample** of the population. However, this sample is particularly large at about 98% of the total population! Information about the remaining 2% has been deduced using statistical techniques.

BBC 1 CATEGORIES TV RADIO COMMUNICATE WHERE I LIVE INDEX SEARCH Go

BBC NEWS

You are in: UK
Monday, 30 September, 2002, 03:40 GMT 04: 40 UK

How was the Census compiled?

By Christine Jeavans
BBC News Online

The Office of National Statistics is confident the 2001 census accurately covers 100% of the UK population – using a new method to fill in the inevitable blanks.

But this has led to accusations officials have "made people up" to complete the statistics.

How do you know a person exists? Do they need a name? Must they really live where you think they do?

Not according to the Office of National Statistics, which says it is so sure that 2% of the population is out there – just not on the census – that it has come up with details including age, sex and addresses for a million or so "individuals".

Statisticians "found" them by looking at the data gathered at the time of the census and working out how many people had been missed out.

They then used advanced statistical techniques to estimate the personal details of each one.

But the number-crunchers strenuously deny that these are figments of the census compilers' imaginations.

"No census anywhere gets everyone to respond – and we haven't," said John Pullinger, Director of Social Statistics and Census.

But "nobody is made up," he insisted.

"These are the real people in the country who we have been unable to find using the census and the census survey."

"Nobody is made up ... these are the real people in the country who we have been unable to find using the census"

John Pullinger
Census director

Completing the picture

88% sent back forms
6% gave over forms on doorstep
Basic info on a further 4% gathered by census takers
2% "imputed" after separate survey

In many instances data is only collected for a sample of a population – for example, most **surveys** into things like voting intentions are based on quite small samples of only just over 1000 individuals. However, because these samples are deliberately designed to be representative of the entire population, taking into account factors such as age, gender, occupation, area of residence and so on, the pollsters can arrive at results that accurately reflect what will happen when everyone votes.

The census is very important because it provides planners, for example, with an accurate picture of the population. They are therefore able to decide how many schools, hospitals and other facilities are needed. You can sec from this **population pyramid** how the population of the UK changed between 1951 and 2001. At a glance it looks as though fewer schools will be needed over the next few years.

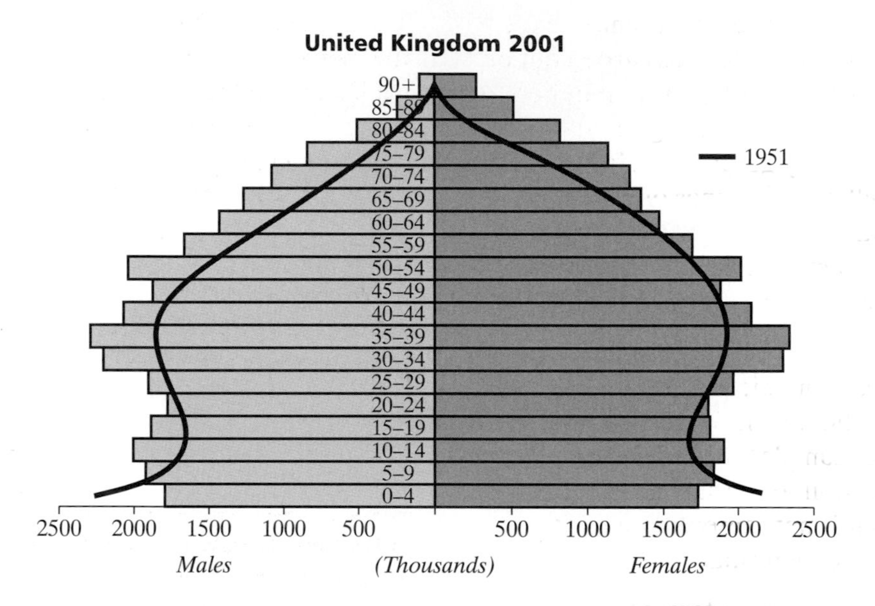

In this chapter you will learn about how to collect data and then how to calculate some common measures of location and spread. You will also learn how to plot some useful charts and graphs to help make sense of data.

1.1 Collecting data

Market research

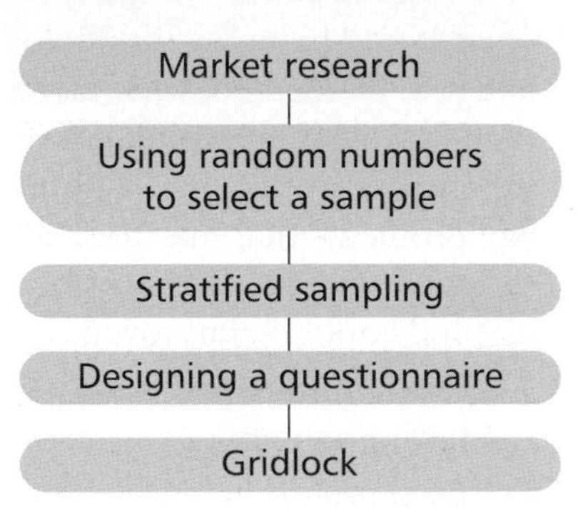

There are many different types of data. Descriptive data, such as gender, colours, types of employee, subjects studied and so on are known as **qualitative data** because it deals with qualities. **Quantitative** data is numerical data. It can be either discrete or continuous. **Discrete** data takes only particular, exact values. Examples include the number of children in a family, shoe sizes and so on. **Continuous** data can take any value in a range. Measurements such as height, weight and temperature fall into this category.

An obvious way of ensuring that data is representative is to collect data from the whole **population**. The government attempts to do this every ten years in a **census**. The last census was carried out on 29 April 2001 when all households in the country were required to complete a questionnaire to give datasets describing the state of the whole nation. These datasets are used by government for policy analysis and in the allocation of billions of pounds of public expenditure. The data are also extremely valuable to commercial organisations and are widely used in market analysis. Information about the 2001 Census is available from the government statistics website.

The **population** consists of all the members of a group being studied, e.g. all students in a college, all cars manufactured by a car company.

Discussion point

In what ways might the UK census fail to get information from the whole population?

In many situations it is impossible, impractical or too expensive to collect data from the whole population. In such cases information should be collected from a **random** sample so that it is **representative** of the whole population. If the sample is not random it may be **biased**. This means that it is not representative of the population and could lead to misleading results.

A **random sample** is a sample in which every member of the population has an equal chance of being included.

There are many ways in which organisations attempt to get information about the population. They might send out questionnaires, employ people to make contact by telephone or stop people in the street to ask for answers to a variety of questions. Data that is collected directly from the population is called **primary** data.

Some organisations collect information about people and then sell it on to other organisations. For example, supermarkets collect information about their customers using the details that they can gather from the scanning of goods at their tills. This information may be sold to other commercial companies that are interested in what the customers are buying. Data that has been collected by someone else is called **secondary** data.

Some methods of collecting information are better than others in giving reliable and representative data.

A market researcher in a shopping mall

Activity 1.1A

A market research company is trying to find out if the people in a town would be in favour of a new sports centre, where they would like it to be located, what facilities they would use and how much they would be prepared to pay. Several methods of carrying out a survey are listed below. Comment on each method, describing any problems that may arise and giving reasons why the sample is not random and is likely to be biased. Suggest ways in which each method could be improved.

1 Phone all the people whose surnames begin with A in the local phone book.
2 Interview people who come out of a local supermarket between 9 am and noon one Monday morning.
3 Send questionnaires by post to all members of local sports clubs.
4 Ask a local radio station to broadcast something about the proposal and invite people to phone or e-mail giving their views.
5 Interview people who arrive at the local railway station between 8 am and 5 pm one Friday.
6 Leave questionnaires in the local library.

Discussion point
How would you go about finding a representative sample?

Using random numbers to select a sample

You can use random numbers to select a random sample. In this activity you will use random numbers to identify a random sample of people from a list who might then be invited to take part in a survey.

Random numbers
You can use a printed sheet of random numbers or your calculator to generate random numbers – see **Book 1 Activity 4.2A**.

Activity 1.1B

Eighty people are listed in the table below.
Use random numbers to select a sample of 20 names.

1	K Abagusii	21	R Daniels	41	L Jaffrey	61	H Rhys
2	H Apps	22	L Darcy	42	S Katsuichi	62	W Riley
3	D Ash	23	S Davies	43	I Khan	63	P Roberts
4	L Atkin	24	F Dean	44	A Lewis	64	S Roche
5	A Binns	25	C Dent	45	B Lovell	65	R Ryan
6	G Bobaljik	26	M Devlin	46	T Major	66	T Scott
7	N Busby	27	B Dodd	47	R Marks	67	B Setna
8	P Cadman	28	N Draper	48	C Mason	68	V Shaw
9	F Cain	29	F Du-Bois	49	J Matsumara	69	P Singh
10	J Cannock	30	K Duffy	50	S Moore	70	J Smith
11	S Carling	31	G Edwards	51	R Naik	71	K Stevens
12	L Carr	32	S Elgar	52	C Norton	72	D Sweeney
13	N Chadd	33	M Evans	53	V O'Neil	73	R Tait
14	K Chen	34	A Farr	54	N Parry	74	L Thomas
15	L Chiu	35	B Ferris	55	M Platt	75	J Thorpe
16	H Clarke	36	K Fox	56	D Pope	76	D Vogul
17	M Clay	38	C George	58	E Prasad	78	G Wadia
18	P Cohen	37	J Grey	57	G Purvis	77	R Waite
19	C Crisp	39	T Hakim	59	S Read	79	A Webb
20	V Czink	40	M Hicks	60	T Reeves	80	S Young

Because there are 80 names, you will need to use pairs of random digits to select the names you require.

For example, the random number 0.3982876544 would be split into pairs of digits: 39, 82, 87, 65, 44, …

The first number, 39, would give a person, T. Hakim, for the sample. The next numbers, 82 and 87 are too large and would be ignored. The next number, 65, would give another person, R. Ryan for the sample, and so on …

Stratified sampling

Sometimes within the population there are sub-groups with different characteristics that will affect whatever is being measured. In such cases a **stratified** sample can be used to give measures that are representative of the whole population.

For example, 55% of the people listed in **Activity 1.1B** are male and the rest female. A representative sample should include the same proportion of males and females.

For a sample of 20:
No. of males = 55% of 20 = $0.55 \times 20 = 11$
No. of females = $0.45 \times 20 = 9$

To select a representative sample you need to list the males and females separately and then randomly select 11 of the males and 9 of the females.

More than one characteristic of the population may be relevant. As well as gender, age is often important. The table below gives estimates, to the nearest hundred, of the population of Gloucester in different gender and age categories.

	Under 16	16–59	Over 60	Total
Male	12 200	31 100	9700	53 000
Female	12 000	31 500	10 900	54 400
Total	24 200	62 600	20 600	107 400

If a representative sample of 1000 is required, the number of males under 16 in the sample should be $\frac{12\,200}{107\,400} \times 1000 = 114$.

A **stratified sample** is obtained by taking samples from each sub-group (stratum) of a population in proportion to the size of the sub-group in the population.

Check point
Total = 11 + 9 = 20

Discussion point
Can you explain this calculation to find the number of males aged under 16 years in a representative sample of 1000?

Activity 1.1C

1 Using the figures given above for the population of Gloucester, calculate the number of people there should be in each of the other categories in a representative sample of size 1000.

2 The Office for National Statistics calculates the Average Earnings Index in Great Britain, based on a sample survey of employers. The sample includes all firms with 1000 or more employees. Firms with 20–99 employees are sampled at approximately 1 in 20; firms with 100–499 employees are sampled at approximately 1 in 4 and firms with 500–999 employees sampled at approximately 1 in 2.

An estimate of the number of firms in each of these categories is given below:

Number of employees	Number of firms
20–99	65 749
100–499	14 213
500–999	3288

Check point
Make sure that you have a total sample size of 1000.

a Calculate the number of firms from each category that are asked to take part in the survey.

b The Office for National Statistics states that '*The sample is not completely representative of the economy…*'
Suggest reasons why they might say this.

4 Imagine that funds have been allocated for a new common room for students in your school or college. You are asked to carry out a survey to ask for views about what should be provided in the common room. Describe in detail how you would select a representative sample of 100 students from the student population.

Discussion point
What student characteristics do you think would be important if a survey was carried out in your school or college about refectory food?

Designing a questionnaire

Questionnaires are often used to collect information and opinions. If you are asked to design a questionnaire it is important to think carefully at the outset about what you wish to find out and how you are going to analyse the answers.

Before the questionnaire is distributed widely it is important to carry out a trial run to see whether there are any problems with the questions and whether it is likely to give the kind of data you need.

Activity 1.1D

Here are some questions from a questionnaire about the use of hair shampoos and conditioners.

Criticise each question and suggest an alternative.

1 What colour of hair do you have?

Blonde ☐ Brown ☐ Black ☐

2 How greasy is your hair? 1 2 3 4 5

3 Do you use a shampoo or conditioner?

Yes ☐ No ☐

4 How often do you wash your hair?

Less than once ☐ Every day ☐

Twice per week ☐ Once per week ☐

5 How much would you pay for a shampoo and conditioner?

Up to £2 ☐ £1–£3 ☐

£3–£5 ☐ Over £5 ☐

Writing a questionnaire

- Questions should not be ambiguous.
- Questions should be short, simple and easy to understand.
- Clear instructions should be given about how the question should be answered.
- Avoid leading questions, i.e. questions that make assumptions or introduce bias.
- Avoid personal or embarrassing questions.
- If giving a list to choose from, make sure all possibilities are included.

Nuffield Resource Starter 'Music'

6 Carefully selected ingredients make conditioners more effective. Which of these ingredients would you like to see as constituent elements?

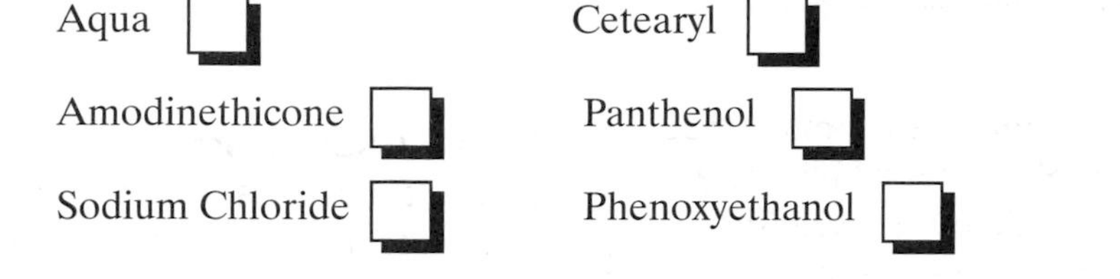

Aqua ☐ Cetearyl ☐

Amodinethicone ☐ Panthenol ☐

Sodium Chloride ☐ Phenoxyethanol ☐

Gridlock

One of the most controversial parts of the London Mayor Ken Livingstone's strategy for tackling the transport crisis in London was the plan to charge for vehicles travelling into the city centre.

Before the strategy was finalised, a draft strategy document was sent to about 1500 organisations and stakeholders (including the 33 London boroughs, 73 London MPs and MEPs, and around 1400 other organisations, including libraries).

In addition, a leaflet was delivered to households in London informing members of the public that they could telephone a dedicated freephone line to request a copy of a 24-page summary document – *Highlights of The Mayor's Draft Transport Strategy* or the full *Draft Strategy*. The Highlights document included a 'pull out' four-page freepost return structured response form. Both the summary and the questionnaire were also made available via the Mayor of London's website. A report was published in July 2001 summarising the results of this consultation. All of this documentation could be downloaded from the Greater London Authority's website.

The Commission for Integrated Transport employed MORI to carry out a survey covering the rest of the country as well as London. The questionnaire and report are available from the Commission for Integrated Transport's website.

Traffic in London

Discussion point

How would you organise a survey to collect representative views from the population of

- London
- the whole country?

MORI (Market and Opinion Research International) is the largest independently owned market research company in the United Kingdom.

Activity 1.1E

Resource Sheet 1.1E

1 Consider carefully the way in which views were sought about the London transport strategy from organisations and members of the general public. Do you think they were likely to give a representative sample? Explain your answer.

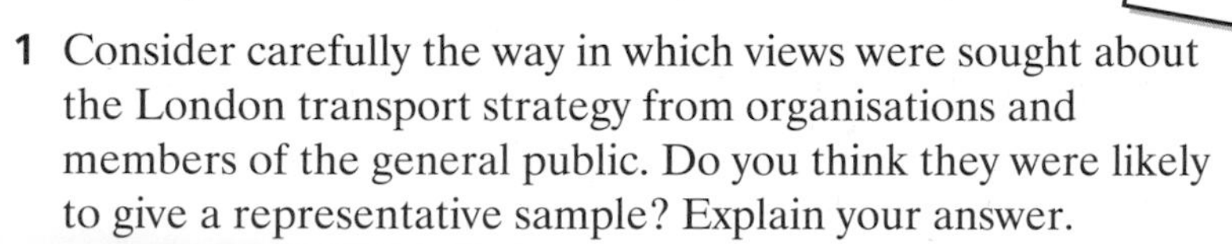

2 Look at the questions included in the MORI questionnaire. Describe any ways in which you think any of them could be improved.

3. Download the MORI report from the Commission for Integrated Transport's website. Use Internet searches to find reports of the survey. Comment on the way in which the results were reported in the official report and in the press.
4. Download *The Mayor's Transport Strategy Report on Public Consultation on the Draft Transport Strategy* from the Greater London Authority's website. Comment on the similarities and differences between this document and the MORI report.

1.2 Measures of location and spread

When summarising a set of quantitative data or comparing one set with another it is often useful to use a measure of location and a measure of spread.

A **measure of location** means an average or typical value. The measures of location most often used are the mean, mode and median.

The **mode** or **modal value** is the most common value.

The **modal group** is the group with the highest frequency – or, if the groups are of different widths, the highest frequency ÷ width.

The **median** is the middle value in an ordered list.

The **mean** is found by dividing the total of the values by the number of values.

A **measure of spread** is a quantity that gives an indication of the variability within a set of data. The measures of spread most often used are the range, interquartile range and standard deviation.

The **range** is the difference between the largest and smallest values.

The **interquartile range** is the difference between the value that is a quarter of the way through the distribution and the value that is three-quarters of the way through the distribution.

The **standard deviation** is the square root of the mean squared deviation from the mean.

You will have met some measures of location and spread before. However, you may not have met standard deviation. If not, don't take fright at this difficult definition – it is not quite as bad as it sounds!

Gooooal!

The question 'Who is best?' arises in many walks of life, not least in the area of sport. Sometimes an attempt is made to answer this question using a single event and sometimes from a series of events using statistics in one way or another.

In the first stage of the World Cup Finals, teams that have qualified play each other in groups of four. The teams with most points from each group then take part in a knock-out competition in the second stage. If the score is a draw at full time in a knock-out game, the teams play extra time. If a team scores a goal in extra time, the match immediately ends and the team that scored this 'Golden Goal' is the winner. If the score is a draw at the end of extra time, the teams take part in a penalty shoot-out.

Suppose you wanted to know which team was the ***best at scoring goals*** in the 2002 World Cup. This may not be Brazil, the team that eventually won the cup. The table that follows shows the number of goals scored by each team in each match. In the next activity you will compare the teams' mode, median and mean scores.

Discussion point

This is just a brief summary of the rules. If you have time, find out what the rules are in more detail. Do you think this is a good way of deciding which is the best team? Changes to the procedures have been suggested, such as widening the goals in extra time. What do you think?

	G1	G2	G3	R	Q	S	F/T
Argentina	1	0	1				
Belgium	2	1	3	0			
Brazil	2	4	5	2	2	1	2
Cameroon	1	1	0				
China	0	0	0				
Costa Rica	2	1	2				
Croatia	0	2	0				
Denmark	2	1	2	0			
Ecuador	0	1	1				
England	1	1	0	3	1		
France	0	0	0				
Germany	8	1	2	1	1	1	0
Ireland	1	1	3	2^A			
Italy	2	1	1	1^B			
Japan	2	1	2	0			
Korea	2	1	1	2^C	5^D	0	2
Mexico	1	2	1	0			
Nigeria	0	1	0				
Paraguay	2	1	3	0			
Poland	0	0	3				
Portugal	2	4	0				
Russia	2	0	2				
Saudi Arabia	0	0	0				
Senegal	1	1	3	2^E	0^F		
Slovenia	1	0	1				
South Africa	2	1	2				
Spain	3	3	3	3^G	3^H		
Sweden	1	2	1	1^I			
Tunisia	0	1	0				
Turkey	1	1	3	1	1^J	0	3
United States	3	1	1	2	0		
Uruguay	1	0	3				

Key

G – group games
R – first round after the group games
Q – quarter-finals
S – semi-finals
F/T – final and third place play-off

Notes

A after penalty shoot-out (after full time 1)
B after extra time (after full time also 1)
C after extra time (after full time 1)
D after penalty shoot-out (after full time 0)
E after extra time (after full time 1)
F after extra time (after full time also 0)
G after penalty shoot-out (after full time 1)
H after penalty shoot-out (after full time 0)
I after extra time (after full time also 1)
J after extra time (after full time 0)

Activity 1.2A

Resource Sheet 1.2A

1 **a** Write down the modal score for each team. List any problems that arise.

b **i** Which team has the highest mode?

ii Do you think this team is the best at scoring goals?

c Do you think the mode gives a 'typical' score for each team? Explain your answer.

The **mode** is the most common value.

Listing Denmark's scores in order of size gives: 0 1 2 2

Middle values

The median is halfway between 1 and 2.

i.e. Median $= \frac{1+2}{2} = 1.5$ goals.

The **median** is the middle value in an ordered list.
(If there are two middle values, the median is halfway between them.)

The **position** of the median is given by $\frac{1}{2}(n + 1)$ where n is the number of values.
For just 4 scores the position is 2.5, i.e. halfway between the 2nd and 3rd values.

Activity 1.2A (continued)

2 a Find the median score for each of the teams.
 b i Which team has the highest median score?
  ii Do you think this team is the best at scoring goals?
 c Do you think the median gives a 'typical' score? Explain your answer.

$$\textbf{Mean} = \frac{\text{sum of values}}{\text{number of values}} = \frac{\Sigma x}{n}$$

where Σx means the sum of the values and n is the number of values.

The **mean**, denoted by $\bar{x}$, is found by dividing the total score by the number of games.

For example, Denmark's mean score was:

$$\bar{x} = \frac{\Sigma x}{n} = \frac{2 + 1 + 2 + 0}{4} = 1.25$$

The mean can also be found using the statistical functions on your calculator. Find out how to enter data and use your calculator to check Denmark's mean score.

Activity 1.2A (continued)

3 a Find the mean score for each of the teams.
 b Do you think Germany's mean score is a typical value?
 c i Do you think Korea's mean score is a typical value?
  ii What special aspects of Korea's games increased their mean score?
 d i Which team has the highest mean score?
  ii Do you think this team is the best at scoring goals?

4 Which average do you think is most useful for telling you the team that is best at scoring goals?

$\bar{x}$ (read as 'x bar') and the Greek letter μ (pronounced 'mew') are both used to represent the mean. Usually μ is used for the theoretical value of the mean whereas $\bar{x}$ is used for the mean of a sample of values.

An unusually high or low value is called an **outlier**. Outliers can seriously affect the size of the mean.

Discussion point
Which scores in the table do you think are outliers? Why did they occur?

Formula 1

In Formula 1 racing the constructors' championship is decided each year by simply adding up the points scored in each race. The scores of the three top constructors between 1995 and 2001 are shown in the table.

	1996	1997	1998	1999	2000	2001	2002
Ferrari	70	102	133	128	170	179	221
McLaren	49	63	156	124	152	102	65
Williams	175	123	38	35	36	80	92

Activity 1.2B

1 Which constructor won the constructors' championship most often during this period?

2 a For each constructor find
   i the mean   ii the median
   number of points.

  b Which of these measures do you think gives the better method of comparing the constructors? Why?

3 Which constructor do you think is the most consistent? How did you decide?

Discussion point
What problem arises if you try to find the mode for each constructor?

Measures of spread are often used in an effort to measure variability. In the following activity you will find measures of spread for the Formula 1 constructors.

The range is the easiest measure of spread to find.

The **range** is the difference between the maximum and minimum scores.

Range = Maximum value − Minimum value
Note: this gives you a numerical value.

Activity 1.2B (continued)

4 a Find the range of scores for each constructor.

  b i Which constructor has the smallest range?
   ii Is this the constructor you thought was most consistent?

The range is obviously affected by unusually high or low scores. For this reason the interquartile range is often used instead. This is the range between the upper and lower quartiles.

The lower quartile is a value one-quarter of the way through the distribution.

The **lower quartile** of n values is given by the $\frac{1}{4}(n + 1)$th value. When there are 7 values, the lower quartile is therefore the 2nd value.

The upper quartile is a value three-quarters of the way through the distribution.

Ferrari's scores in order of size are:

70	102	128	133	170	179	221
	↑ Lower Quartile		↑ Median		↑ Upper Quartile	

Ferrari's interquartile range (IQR) = 170 – 102 = 68.

The **upper quartile** of n values is given by the $\frac{3}{4}(n + 1)$th value. When there are 7 values, the upper quartile is given by the 6th value.

Interquartile range = Upper quartile − Lower quartile

Note: if there were just 6 values the position of the lower quartile (LQ) would be given by:

$\frac{1}{4}(6 + 1) = \frac{7}{4} = 1\frac{3}{4}$

This means the LQ is three-quarters of the way between the 1st and 2nd values.

Using the first 6 values only, the LQ is three-quarters of the way between 70 and 102.

$LQ = 70 + \frac{3}{4} \times (102 - 70) = 94$

Activity 1.2B (continued)

5 a Find the interquartile range of scores for each of the constructors.

b Which constructor has

i the smallest interquartile range

ii the largest interquartile range?

c On this basis which is the most consistent constructor? Compare your answer with your answer to **question 4b**.

Standard deviation is a measure of spread that uses ***all*** the values. It is used more widely in statistical analysis than the range or interquartile range. Unfortunately it is not so easy to find or understand.

Imagine the values in a data set are arranged along a scale as shown below. The mean, $\bar{x}$, will lie somewhere near the centre.

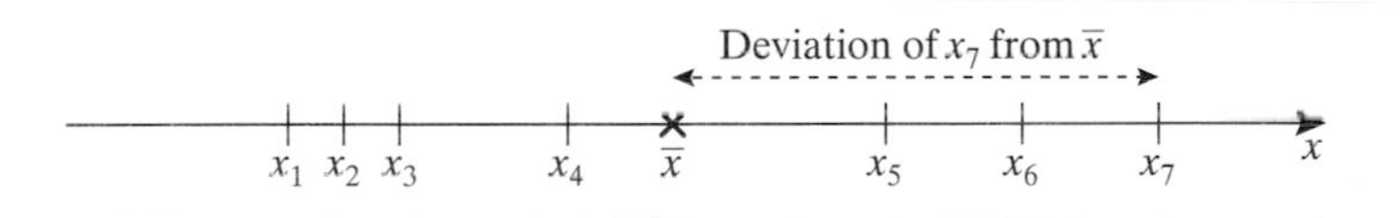

The **deviation** of a value from the mean is the difference between $\bar{x}$ and that value.

One idea for measuring spread in a set of numbers is to find the mean deviation from the mean.

The next part of this activity shows what happens if you take the mean from each value and then find the mean of these differences.

Activity 1.2B (continued)

6 **a** Copy Ferrari's scores onto a spreadsheet as shown below.

b Use the formulae shown at the bottom of column B to find the mean.

	A	B	C	D
1		x	**x-mean**	**(x-mean)²**
2	1996	70	=B2-\$B\$10	=C2^2
3	1997	102	=B3-\$B\$10	=C3^2
4	1998	133	=B4-\$B\$10	=C4^2
5	1999	128	=B5-\$B\$10	=C5^2
6	2000	170	=B6-\$B\$10	=C6^2
7	2001	179	=B7-\$B\$10	=C7^2
8	2002	221	=B8-\$B\$10	=C8^2
9	**Total**	=SUM(B2:B8)	=SUM(C2:C8)	=SUM(D2:D8)
10	**Mean**	=B9/7	=C9/7	=D9/7

c Use the formulae shown in column C to find

i the difference between each value and the mean

ii the total of these differences

iii the mean of the differences.

Why is the mean difference not a useful measure of spread?

d Use the formulae shown in column D to find

i the square of each of the differences

ii the total of these squared differences

iii the mean squared difference.

Would the mean squared difference be a better measure of spread?

Discussion point
Why is squaring the differences helpful?

The problem with the mean of the differences is that sometimes the difference, $x - \bar{x}$, is positive and sometimes it is negative. As you have seen, these deviations 'cancel' each other out.

However, squaring them always gives positive values.

The mean squared deviation from the mean, $\frac{\Sigma(x-\bar{x})^2}{n}$, is a measure of spread called the **variance**.

The **standard deviation** is the square root of the variance,

i.e. $s = \sqrt{\frac{\Sigma(x-\bar{x})^2}{n}}$.

This formula can be written in another form as $s = \sqrt{\frac{\Sigma x^2}{n} - \bar{x}^2}$.

Find the square root of the value in cell D10 to give the standard deviation of Ferrari's scores. You should find that it is 47.0 (to one decimal point).

The **standard deviation** is the square root of the mean squared deviation from the mean.

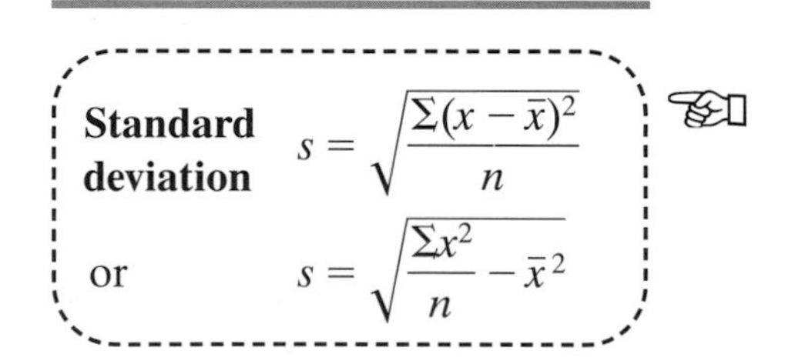

Discussion point
When $\bar{x}$ is not a whole number, the latter form is often more convenient. Can you see why?

Calculators usually have two keys for standard deviation, marked by σ_n and σ_{n-1}. The σ_n key gives the standard deviation of a sample or of a whole population when all of the values are entered. The σ_{n-1} key gives an estimate of the standard deviation of a population based on the values in a sample. The values given by the two keys are usually very similar. For the time-being use the key marked by σ_n. Use your calculator now to check the standard deviation of Ferrari's scores.

s and the Greek letter σ (pronounced 'sigma') are both used to represent standard deviation. Usually σ is used for the theoretical value of the standard deviation whereas s is used for the standard deviation of a sample of values.

Activity 1.2B (continued)

7 **a** Use your spreadsheet to find the standard deviation of the scores for each of the other constructors.

b Check your answers using your calculator.

c Use the standard deviations you have found to suggest which of the constructors has

i the most variable scores

ii the most consistent scores.

d Compare your answers with **questions 4 and 5**.

Share the work in **8a** with other students if you wish and pool your results.

8 **a** For each of the teams in the World Cup table in **Activity 1.2A** find

i the range

ii the interquartile range

iii the standard deviation.

b Which of these measures give a fair way of comparing how consistent the teams are at scoring goals? Explain your answer.

Units
Note that the units of the median, mode, mean, range, interquartile range and standard deviation are all the same as the units of the variable under consideration.

Airport delays

Data about flight delays is given in the table below. It includes all flights from the airports at Birmingham, Manchester and Newcastle during August 2001. The flights have been grouped according to the length (to the nearest minute) of any delay.

	Delay (min)					
Airport	**Early–15**	**16–30**	**31–60**	**61–180**	**181–360**	**Total**
Birmingham	8291	1155	630	420	0	10 496
Manchester	12 876	2176	1632	1269	181	18 134
Newcastle	3549	404	315	180	45	4493

You can compare the punctuality of flights from the three airports using the mean to give an average for each airport and the standard deviation as a measure of spread.

The data for Birmingham is shown again in the **frequency table** below.

Delay (min)	No. of Flights
Early–15	8291
16–30	1155
31–60	630
61–180	420
181–360	0
Total	10 496

Frequency is the number of items in a group (class).

The first group includes flights with delays up to 15.5 minutes, the second group includes flights with delays from 15.5 minutes up to 30.5 minutes and so on. The CAA counts each early flight as a zero delay. This means the **class boundaries** are at 0, 15.5, 30.5, 60.5, 180.5 and 360.5 minutes.

To estimate the mean, each flight delay is taken to be at the mid-value of its group, i.e. halfway between its boundaries.

For example, the mid-value of the first group is $\frac{0 + 15.5}{2} = 7.75$ min.

So all 8291 flights in the first group are taken to have been delayed by 7.75 minutes, giving a total delay for that group of $7.75 \times 8291 = 64\,255.25$ minutes.

Discussion points

It is important that you are able to identify class boundaries correctly. Can you explain why in this case these are at 0, 15.5, 30.5, etc.?

Why is it usual to use the value at the midpoint of each group when you calculate an estimate of the mean for grouped data?

Mean $\bar{x} = \frac{\Sigma xf}{n}$

where n is the total frequency, x denotes the mid-value of a group and f its frequency.

You can find the total delay for each group using a spreadsheet. Use the formulae in the spreadsheet below to do this and then find estimates of the mean and standard deviation.

	A	B	C	D	E
1	Delay (min)	Mid-interval value x (min)	f	xf	x^2f
2	Early–15	7.75	8291	=B2*C2	=B2*D2
3	16–30	23	1155	=B3*C3	=B3*D3
4	31–60	45.5	630	=B4*C4	=B4*D4
5	61–180	120.5	420	=B5*C5	=B5*D5
6	181–360	270.5	0	=B6*C6	=B6*D6
7		Total	=SUM(C2:C6)	=SUM(D2:D6)	=SUM(E2:E6)
8					
9	mean	=D7/C7			
10	sd	=SQRT(E7/C7-B9^2)			

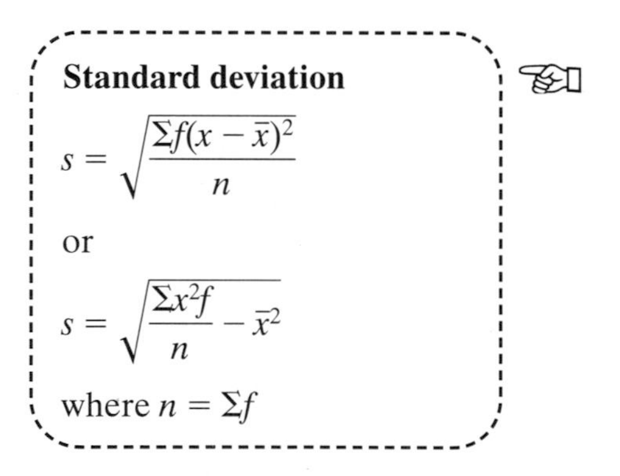

Standard deviation

$$s = \sqrt{\frac{\Sigma f(x - \bar{x})^2}{n}}$$

or

$$s = \sqrt{\frac{\Sigma x^2 f}{n} - \bar{x}^2}$$

where $n = \Sigma f$

You can also find the mean and standard deviation using the statistical functions on a calculator. After entering the values given in the x and f columns of the table, the calculator will give the value of the mean, $\bar{x}$, and two values for the standard deviation, σ_n and σ_{n-1}. The calculator will also give the total of each column, usually denoted by n, Σx and Σx^2. Find out how your calculator works and use it to check the values you found with your spreadsheet.

Discussion point

The mean and standard deviation calculated here are estimates not accurate values. Why is this?

On your calculator use σ_n for standard deviation.

Activity 1.2C

1 a Use a spreadsheet to find the mean and standard deviation of the flight delays at
  i Manchester ii Newcastle.

 b Use the statistical functions on your calculator to check your answers to **part a**.

 c Compare the mean delay times for the three airports.
  i Which airport had the greatest average delay time?
  ii Which airport had the least average delay time?

 d Compare the standard deviation of the delay times for the three airports.
  i At which airport were the delay times most variable?
  ii At which airport were the delay times most consistent?

2 When values are grouped, it is not possible to find the mode. However, it is possible to identify a modal group, i.e. the group with the highest frequency.
What is the modal group for each airport?

3 Is it possible to find the range of the delay times for each airport? Explain your answer.

Since there were 10 496 flights from Birmingham, the position of the median is halfway between the 5248th and the 5249th delay times in an ordered list.

As the first group (i.e. Early to 15 minutes late) contains 8291 times, the median lies somewhere in this group.

If we assumed that the delay times in this group were **evenly spread** between 0 and 15.5 minutes we could estimate the median time to be $\frac{5248.5}{8291} \times 15.5 = 9.8$ minutes.

4 a It is not really appropriate to make this assumption in this situation. Why not?

 b Find the ***position*** of the median delay time and the group in which it lies for flights from
  i Manchester ii Newcastle.

 c Find an estimate of the median delay time for flights from
  i Manchester ii Newcastle.

Practice Sheet: Measures of location and spread

The **modal group** is the group with the highest frequency.

The median is the $\frac{1}{2}(n + 1)$th value where n is the number of values.

This method is called **linear interpolation**.

Wage negotiations

This section includes using a graphical method to estimate the median and quartiles of a grouped frequency distribution.

During wage negotiations both employers and unions use statistics to support their case. Sometimes unions may not have access to precise information. This was the case for one union when they were only given wage data in bands of £5000. The union said that this made it impossible to work out the average wage.

The following activity shows how grouping figures in different ways can affect the values found for measures of location and spread. Use the resource sheet and a computer (or your calculator) to complete this activity.

Resource Sheet 1.2D

Activity 1.2D

The table below gives the salaries of a group of workers.

£24 062	£25 950	£19 862	£25 080	£23 508	£34 182
£33 618	£24 950	£36 948	£33 975	£29 037	£19 450
£19 862	£28 490	£34 015	£27 556	£24 860	£28 321
£24 675	£23 932	£28 642	£33 526	£24 984	£19 298
£34 783	£23 904	£37 840	£19 427	£34 881	£23 842
£18 930	£30 814	£26 420	£20 840	£24 390	£21 920
£27 762	£18 930	£23 325	£28 850	£29 960	£23 586
£21 642	£27 359	£38 245	£18 735	£19 329	£26 844
£24 967	£26 995	£29 812	£27 305	£24 381	£24 950
£19 437	£18 930	£24 984	£32 294	£22 372	£19 553

This is **raw data** – it has not been organised in any way.

1 Use the raw data to find accurate values for
- **a** the mean
- **b** the mode
- **c** the median
- **d** the standard deviation
- **e** the interquartile range of the salaries.

Putting the raw data into ascending order on a spreadsheet makes finding the mode and median as well as grouping the data easier.

2 **a** Complete the following table showing the salaries to the nearest £5000.

Salary (£000s)	Frequency
20	
25	
30	
35	
40	
Total	

b What is the modal group of salaries?

c Use the table to estimate
- **i** the mean
- **ii** the standard deviation
- **iii** the median
- **iv** the interquartile range of the salaries.

Discussion point
The first group includes all salaries from £17 500 up to (but not including) £22 500.
Can you explain this?

3 **a** Complete the following table by grouping the salaries in a different way into bands of £5000.

Salary (£000s)	Frequency
15–	
20–	
25–	
30–	
35–	
Total	

b What is the modal group in this table?

c Use the table to estimate
- **i** the mean
- **ii** the standard deviation
- **iii** the median
- **iv** the interquartile range of the salaries.

Discussion point
In this table the first group includes all salaries from £15 000 up to (but not including) £20 000.
Can you explain this?

4 **a** Compare and comment on the answers you found for the measures of location and spread in **questions 1 to 3**.

b Do you think the union were justified in their complaint?

Discussion point
Which of the three sets of figures found in this activity would you use if you were a manager arguing against a pay rise? What if you were a union leader arguing for a pay rise?

1.3 Statistical charts and graphs

Road safety

'*A picture's worth a thousand words*' was a slogan dreamt up by Frederick Barnard who was the manager of the Street Railways Advertising Company in America in the 1920s. This statement is also true in statistics where charts and graphs can be used to illustrate complex sets of data in ways that make them easy to understand.

This activity uses charts and graphs from road safety reports carried out in 1998.

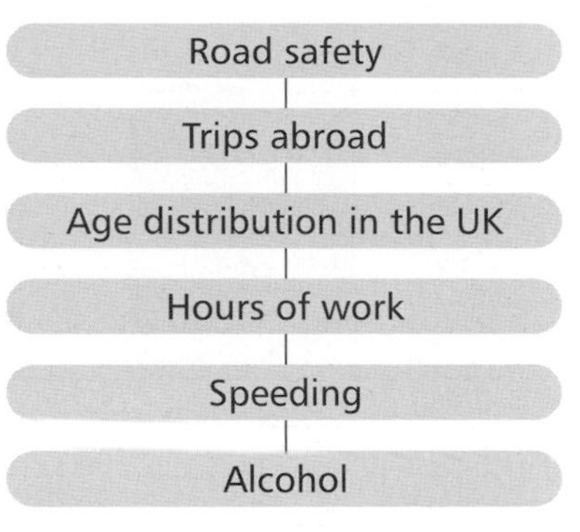

Activity 1.3A

1 The bar chart shows child fatality rates in European countries.

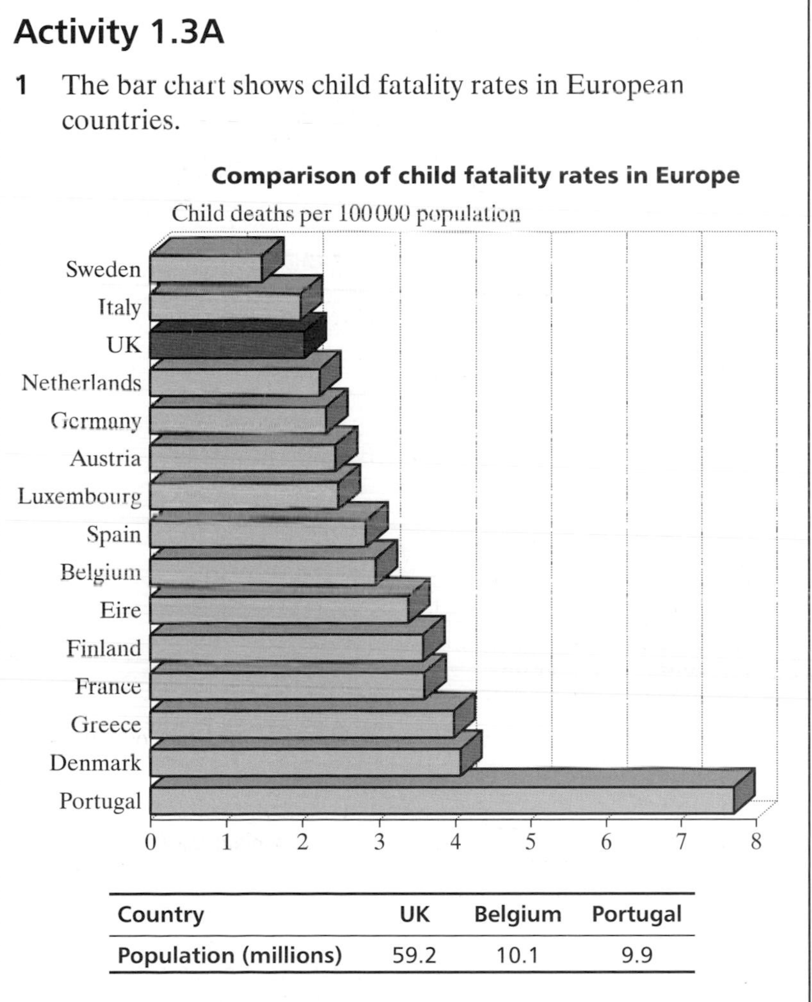

Country	UK	Belgium	Portugal
Population (millions)	59.2	10.1	9.9

a Estimate the number of child fatalities in the UK, Belgium and Portugal. Give your answers in a table.

b Illustrate your results using a bar chart.

c Write a sentence or two describing what you have found.

Bar Charts are usually used to illustrate **discrete** data. The bars can be horizontal or vertical. Gaps are usually left between each bar and the next.

This example is a three dimensional bar chart. Note how difficult it is to read off values accurately. A two dimensional chart is usually preferable.

Discussion points

Why does the bar chart give the number of child deaths per 100 000 population?

It would be better to give the number of child deaths per 100 000 children. Why?

2

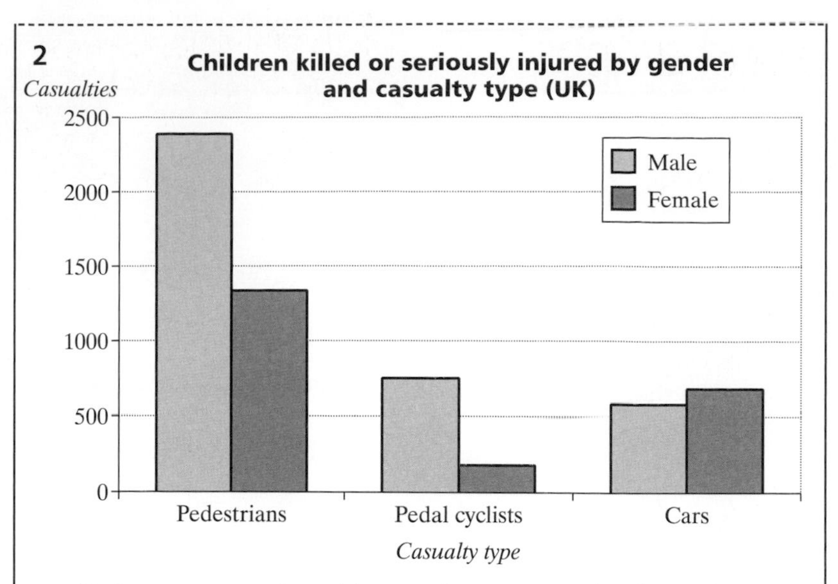

Before you complete this question, answer the discussion points. Imagine you are a journalist who has been given the above comparative bar chart. Write a brief article summarising the information. Choose a headline to go with your article.

3

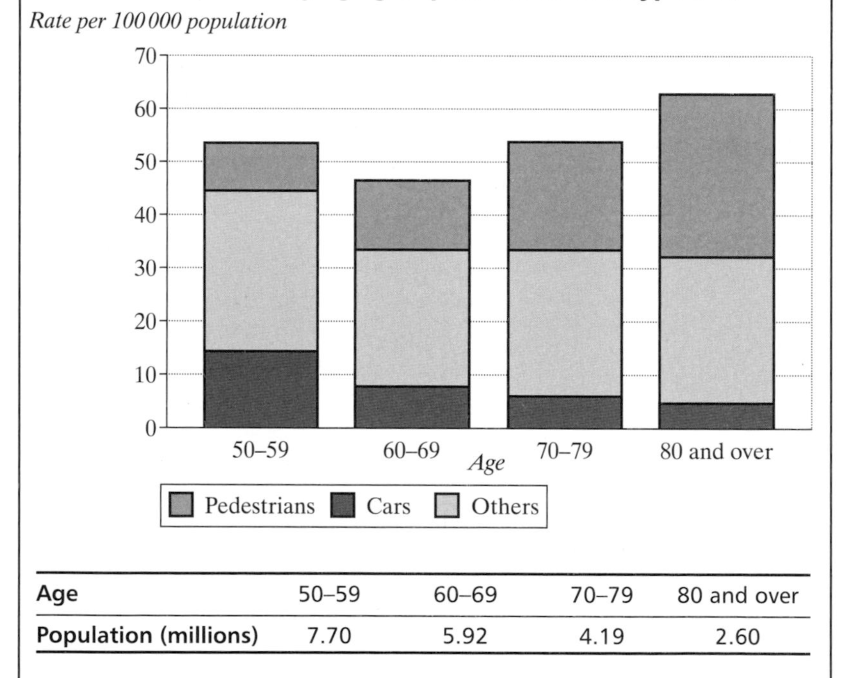

Age	50–59	60–69	70–79	80 and over
Population (millions)	7.70	5.92	4.19	2.60

a Calculate an estimate of the number of casualties in each category within each age group who were killed or seriously injured.

b Write a paragraph or two that compares and comments on the casualty **rates** for the different age groups in each of the three sub-categories.

Comparative bar charts use adjacent bars to compare two or more sub-categories.

In this case there are two sub-categories (male and female) within each casualty type.

Discussion points

Discuss whether the following statements are true, false or you don't know (and if this is the case what additional information you would need to help you decide).

'There are fewer girl cyclists than boy cyclists.'

'Girl pedestrians are much safer than boys.'

'More boys travel by cycle than by car.'

Component bar charts have bars that are split into two or more parts representing different sub-categories.

Component bar charts are used in preference to comparative bar charts when the totals are important.

When this type of chart is used to show percentages within each group the columns all have the same height representing a total of 100% for each group.

Discussion point

Which modes of transport would you expect to be included in the 'Others' category?

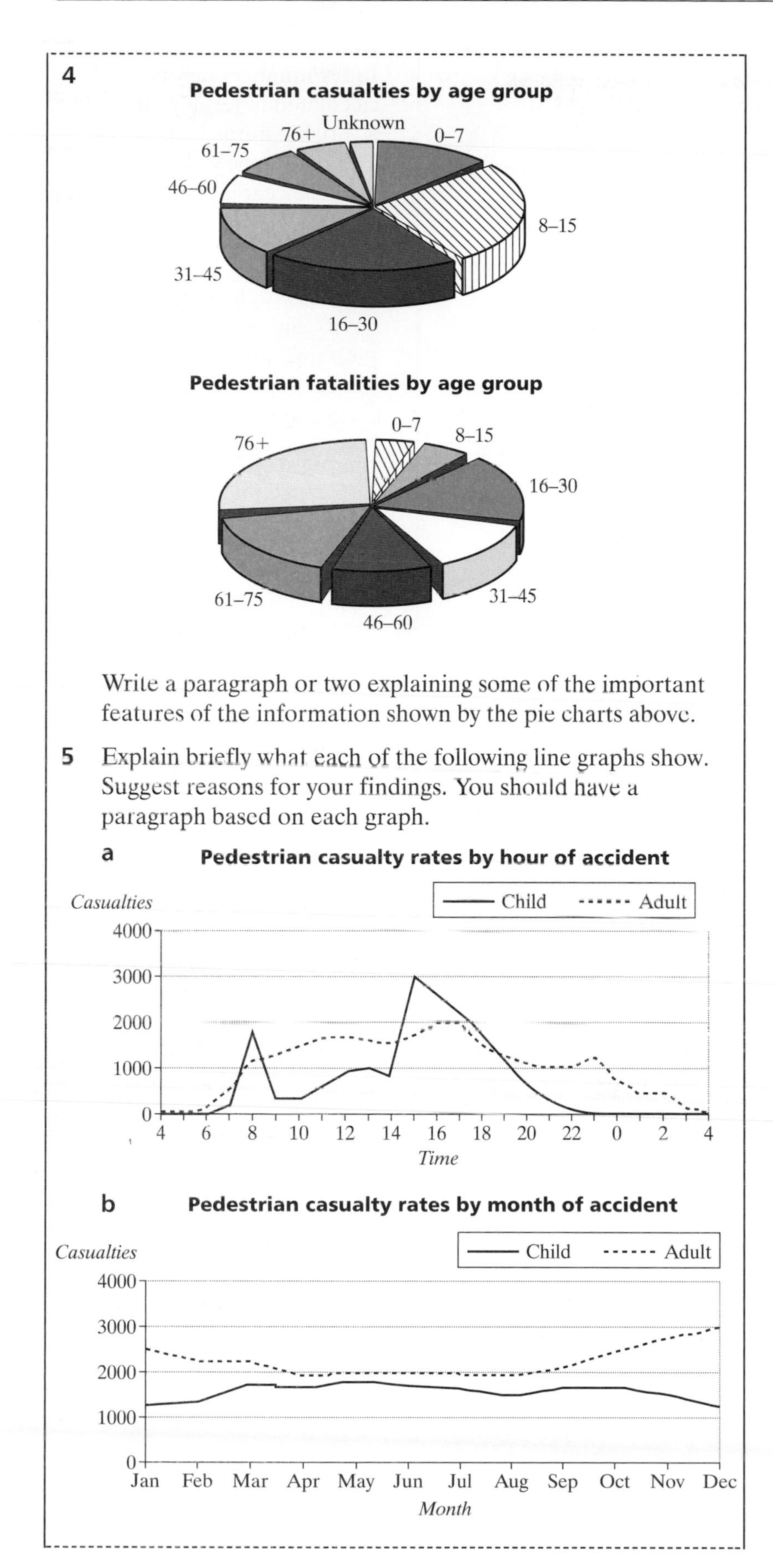

4 Write a paragraph or two explaining some of the important features of the information shown by the pie charts above.

5 Explain briefly what each of the following line graphs show. Suggest reasons for your findings. You should have a paragraph based on each graph.

a Pedestrian casualty rates by hour of accident

b Pedestrian casualty rates by month of accident

Pie charts show how a total is split into parts. Each sector represents a proportion of the total. If the total is known, then the number in each category can be calculated from the angles of the pie chart.

Discussion points

In what ways may 3-D pie charts be misleading?

Why is it not possible to calculate percentages from these 3-D pie charts?

Line graphs are often used to illustrate how a quantity varies with time. Time is shown on the horizontal axis. More than one category can be shown on the same graph.

Discussion points

These line graphs are formed by plotting 24 discrete points and joining each to the next with a line segment. Does the line between the points have any meaning? Why is it plotted?

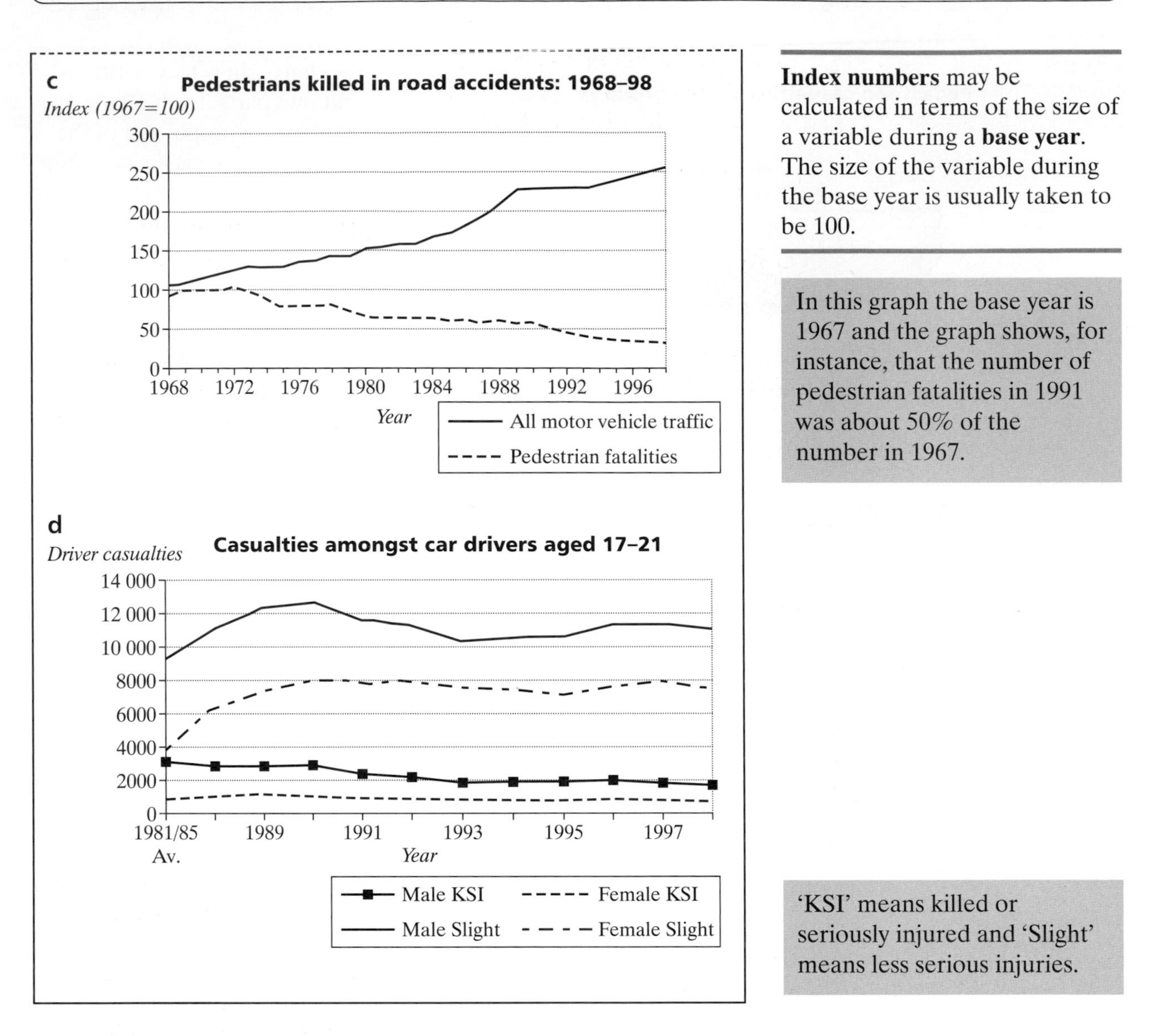

Index numbers may be calculated in terms of the size of a variable during a **base year**. The size of the variable during the base year is usually taken to be 100.

In this graph the base year is 1967 and the graph shows, for instance, that the number of pedestrian fatalities in 1991 was about 50% of the number in 1967.

'KSI' means killed or seriously injured and 'Slight' means less serious injuries.

Trips abroad

The table shows the number of visits, length of stay and amount spent by UK residents when travelling abroad on holiday and business trips.

	No. of visits (thousands)		No. of nights (thousands)		Spending (£millions)	
	Holiday	Business	Holiday	Business	Holiday	Business
North America	3052	965	44 189	8326	2512	1017
EU Europe	26 768	6420	235 886	25 928	9067	2390
Non EU Europe	3193	778	35 599	4298	1290	404
Other countries	3671	709	70 585	10 097	2915	922

Source: Travel Trends 2000 (Crown Copyright 2001)

The pie charts show the way in which the number of holiday and business visits was divided between different regions of the world. But why is the second pie chart so much smaller than the first?

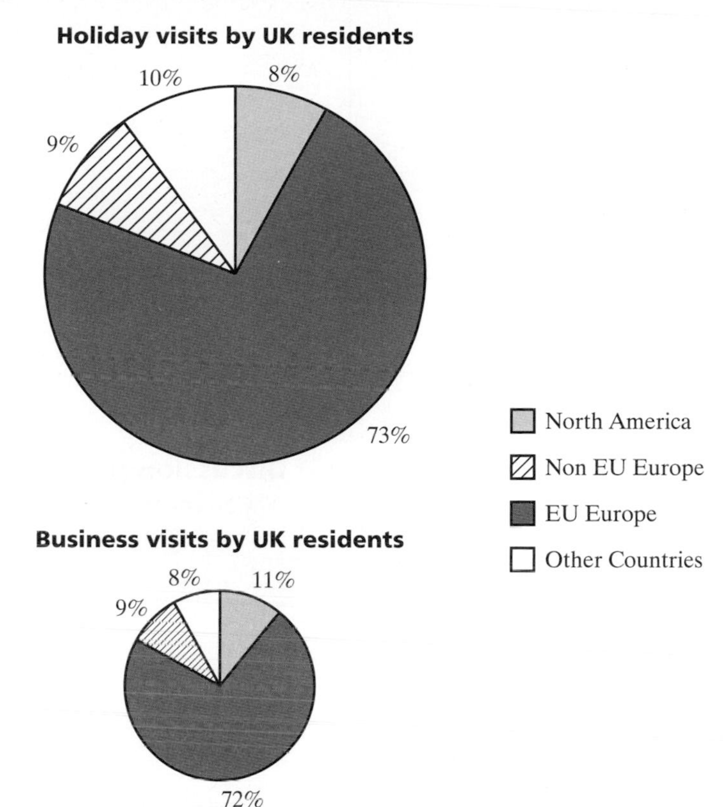

Discussion points

How have the percentages and the angles of the pie charts been calculated?

What are the radii of the two pie charts? How many times bigger is the first pie chart than the second? Why?

The answer to this question is that there were far more visits made abroad for holidays than for business. If you use the figures in the table to calculate the total number of holiday visits and business visits you should find that there were 36 684 thousand holiday visits but only 8872 thousand business visits.

Since 36 684 ÷ 8872 = 4.135, there were about four times as many holiday visits as business visits. In a pie chart **area represents frequency**, so the area of the first pie chart should be about four times the area of the second. This means that the radius of the first pie chart should be about twice as big as that of the second pie chart. Can you explain why?

Comparative pie charts

If the totals represented by two pie charts are n_1 and n_2 then the relationship between their radii is given by:

$$\frac{\pi r_2^2}{\pi r_1^2} = \frac{n_2}{n_1}$$

$$r_2 = \sqrt{\frac{n_2}{n_1}} \times r_1$$

Activity 1.3B

1 Write a few sentences interpreting the holiday and business visits pie charts.

2 **a** Draw comparative pie charts to illustrate the number of nights spent by UK residents when travelling in different regions of the world on holiday and business trips.

b Write a few sentences interpreting the pie charts you have drawn in **part a**.

3 **a** Draw comparative pie charts to illustrate the amount spent by UK residents when travelling in different regions of the world on holiday and business trips.

b Write a few sentences interpreting the pie charts you have drawn in **part a**.

4 Use the information from this activity to write a short newspaper article about travelling abroad. Use the table, your pie charts and/or other statistical charts and calculations to illustrate the points you make.

Age distribution in the UK

The table below shows government estimates of the population of the UK in 2001 and how the population was distributed between different age groups.

	A	B
1	**Age (years)**	**UK Population (thousands)**
2	0–4	3299
3	5–15	7078
4	16–44	21 775
5	45–64	16 316
6	65–74	6239
7	75 and over	5219
8	**Total**	**59 926**

You can use a **histogram** to illustrate **continuous** data that is **grouped**. When drawing a histogram always take care to identify the class boundaries correctly.

For the population data above, the first class is given as 0–4. Because we say a child is 4 right up to the day before their fifth birthday, this class includes all children who are younger than 5 years old. The lower class boundary is 0, and the upper class boundary is 5. The 'width' of the first class is the difference between its boundaries, i.e. 5.

Another problem is that no upper boundary is given for the last group. Often when this occurs it is reasonable to assume that the last group has the same width as the previous group, but in this case that would give an upper boundary of 84. This is obviously too low as many people live beyond this age. In this case it would be more reasonable to use an upper boundary of 100.

Even after sorting out these problems, it is easy to go wrong. Work through the following activity to see what happens if you make the worst mistake of all!

Discussion point

Why do you think the population has been divided into these particular groups?

Age data

Be very careful with class boundaries. Remember that, for example, 0–4 means from 0 up to 5 years old.

Class width
= Upper class boundary – Lower class boundary

Open-ended groups

Choose a sensible boundary for an open-ended group. Where appropriate, assume the group is equal in width to the adjacent group.

Activity 1.3C

1 a Draw a horizontal axis showing ages from 0 to 100 and a vertical axis showing UK Population (thousands) from 0 to 22 000 (thousand).

b Draw bars (without gaps) to show the data given in the table above. Remember to take care with the boundaries.

c The table shows that there are approximately 3 times as many people in the 16–44 group as there are in the 5–15 group. Do you think that the chart you have just drawn gives a true representation of this?

d Compare the number of people in other groups and what your chart shows.

2 a Assuming that the people in each age group in the table are evenly spread between the ages covered by the group, estimate the number of people of each age from 0 years to 99 years. Give your answers in a table.

b Compare your answers to **part a** with your chart.

Note for question 2a

For example, in the first group by assuming that there are equal numbers of 4-year-old children, 3-year-old children, 2-year-old children, 1-year-old children and babies less than 1 year old the number of children of each of these ages is estimated to be: 3299 thousand ÷ 5 = 660 thousand (nearest thousand), i.e. (frequency in group) ÷ (class width).

The chart you have drawn in **Activity 1.3C** gives the wrong impression because the groups have unequal class widths. It is wrong to use frequency (in this case UK population in thousands) on the vertical axis in such cases. A better representation of the data is given by using a histogram based on **frequency density** instead. Frequency density is found by dividing the frequency of each class by its width. This means that **frequency is represented by the area of the bar,** rather than its height.

In a **histogram** a bar is drawn for each group such that the **area** of the bar represents the frequency.

Drawing a histogram

Use frequency density, rather than frequency on the vertical axis. This is essential when the class widths are unequal.

Frequency density

$$= \frac{\text{frequency}}{\text{class width}}$$

The population data is repeated in the table below. It also shows the boundaries, class width and frequency density of some of the groups.

Age (years)	Frequency (thousands)	Lower boundary	Upper boundary	Class width	Frequency density (thousands per year)
0–4	3299	0	5	5	3299/5 = 660
5–15	7078	5	16	11	7078/11 = 643
16–44	21 775	16	45	29	
45–64	16 316				
65–74	6239				
75–99	5219				

A histogram has been started using the completed rows of the table.

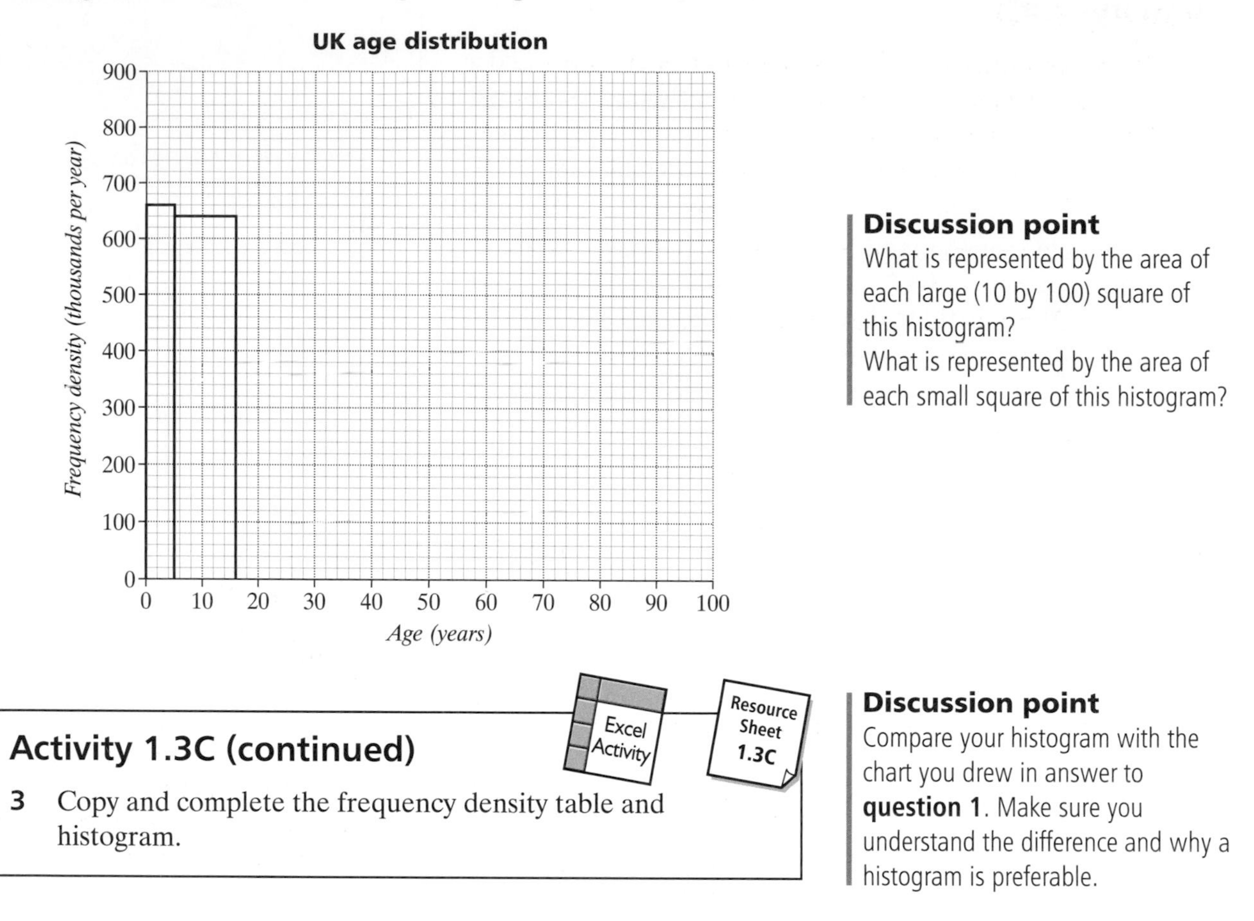

Discussion point

What is represented by the area of each large (10 by 100) square of this histogram?

What is represented by the area of each small square of this histogram?

Activity 1.3C (continued)

Excel Activity

Resource Sheet 1.3C

3 Copy and complete the frequency density table and histogram.

Discussion point

Compare your histogram with the chart you drew in answer to **question 1**. Make sure you understand the difference and why a histogram is preferable.

Hours of work

Surveys are regularly carried out to find the usual weekly hours of workers in the UK. The table shows the results of a survey carried out between February and April 2002.

	Workers (thousands)	
Hours of work	Men	Women
Less than 6 hours	100	305
6–15 hours	519	1555
16–30 hours	941	3731
31–45 hours	8702	5904
Over 45 hours	5381	1333

Discussion point

What do you think are sensible values to use as the lower boundary of the first class and the upper boundary of the last class?

Note that there are gaps between some of the groups. When identifying the class boundaries you need to assume, for instance, that any time between 15 and 16 hours has been rounded to the nearest hour. This suggests the upper boundary of the second group and the lower boundary of the third group are both 15.5 hours. Deal with other gaps between groups in the same way.

There should be no gaps between the bars of a histogram.

Activity 1.3D

1. Draw two histograms, one for men and one for women, to illustrate the data. Use the same scales on your histograms so that they can be easily compared. You may find it useful to draw the two histograms on one piece of graph paper, one above the other.
2. By comparing your histograms, write a paragraph to describe the similarities and differences between the usual weekly hours of work of men and women in the UK.

Speeding

A survey was carried out in 1999 to find out the speeds at which motor vehicles travel on a variety of roads. The table below shows the results for urban roads with a speed limit of 40 miles per hour.

Speed (mph)	Cars	Motorbikes	Light goods	Buses	HGVs
Under 20	24 270	500	1830	300	1940
20–30	89 000	750	8540	2160	8930
30–35	210 260	750	17 995	1860	14 250
35–40	274 690	1100	19 825	1260	13 920
40–45	145 320	800	9300	360	4620
45–50	48 360	750	2440	60	1080
50–60	16 100	400	610	0	300
60 and over	0	250	0	0	300
Total	808 000	5300	60 540	6000	45 340

Discussion point

Here it seems reasonable to start at 0 mph and finish at 70 mph. Other decisions are also acceptable – what would you have chosen and why?

The way in which the speeds are grouped in this table is ambiguous. For example, into which category would a car travelling at 30 mph fall? From this table you do not know. If you are collecting your own data, you should decide and make it clear in your table.
A less ambiguous table is shown below. Can you now see into which category a car travelling at 30 mph would fall?

Speed x (mph)	Cars	Motorbikes	Light goods	Buses	HGVs
$0 \leq x < 20$	24 270	500	1830	300	1940
$20 \leq x < 30$	89 000	750	8540	2160	8930
$30 \leq x < 35$	210 260	750	17 995	1860	14 250
$35 \leq x < 40$	274 690	1100	19 825	1260	13 920
$40 \leq x < 45$	145 320	800	9300	360	4620
$45 \leq x < 50$	48 360	750	2440	60	1080
$50 \leq x < 60$	16 100	400	610	0	300
$60 \leq x < 70$	0	250	0	0	300
Total	80 800	5300	60 540	6000	45 340

It has been assumed that the lower boundary of the first group is 0 mph, the upper boundary of the last group is 70 mph and each group includes its lower boundary, but not its upper boundary. An alternative way of showing this is to label the groups 0– , 20– , 30– , …, 60–(70).

This speed data can be illustrated using a cumulative frequency graph. A cumulative frequency table and graph for the car speed data have been started below.

Speed (mph)	Cars (C.F.)	
<20	24 270	
<30	113 270	i.e. 24 270 + 89 000
<35	323 530	i.e. 113 270 + 210 260
<40	598 220	
<45		
<50		
<60		
<70		

The **cumulative frequency** of a value of the variable is the number of readings that are **less than** (or less than or equal to) this value.

A cumulative frequency table is drawn up by keeping a 'running total' of the frequencies.

The final cumulative frequency should be equal to the total number of readings.

Cumulative frequency graph showing car speeds on 40 mph urban roads

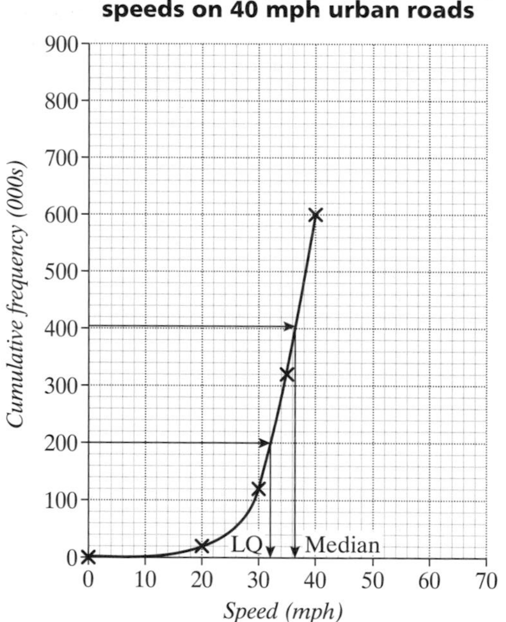

On the graph, a point has been plotted at (0, 0) indicating that no car travelled at less than 0 mph.

Other points have been plotted using the values given in the cumulative frequency table. Note that a point is plotted at the **upper** boundary of each of the original groups.

You can use the cumulative frequency graph to estimate the median and quartiles of the data.
The median speed is the speed that a car halfway through the distribution is driven at.

Here, there are 808 000 readings altogether.
Median ≈ 404 000th reading ≈36.5 mph
LQ ≈ 202 000th reading ≈32.5 mph

The position of the **median** is given by $\frac{1}{2}(n + 1)$ where n is the number of values. If n is large, $\frac{1}{2}n$ is near enough.

The **lower quartile** of n values is given by the $\frac{1}{4}(n + 1)$th value.

The **upper quartile** of n values is given by the $\frac{3}{4}(n + 1)$th value.

Activity 1.3E

Excel Activity

Resource Sheet 1.3E

1 **a** Complete the cumulative frequency table and graph for car speeds.

b Find the upper quartile and interquartile range.

c Use your cumulative frequency graph to estimate the percentage of cars that were breaking the 40 mph speed limit.

Share the rest of this work with other students to save time.

2 Draw cumulative frequency graphs for the speeds of the other types of vehicles. For each type of vehicle find the median and interquartile range and the percentage breaking the speed limit.

3 Compare and interpret the results for the different vehicles. Write a paragraph in which you describe the main differences.

Remember:
Interquartile range = upper quartile − lower quartile

The median and quartiles divide the distribution into quarters. **Percentiles** divide the distribution into hundredths. In this case there are 808 000 readings. The position of the 10th percentile is given by 10% of 808 000, i.e. 80 800.

Look at your cumulative frequency graph of car speeds. Draw a line from 80.8 thousand on the cumulative frequency axis to find the 10th percentile on the speed axis. You should find your answer is approximately 27.5 mph. This suggests that 10% of the cars are travelling at less than 27.5 mph. This also suggests that 90% of the cars are travelling at 27.5 mph or more.

The **10th percentile** is the value that is 10% of the way through the distribution.
The **90th percentile** is the value that is 90% of the way through the distribution.

Percentiles are often used by manufacturers. For example, a furniture manufacturer might use the 5th and 95th percentiles of measurements taken from a sample of people when designing an adjustable chair. In this way the manufacturer can design the chair to suit the middle 90% of the population.

Scientists also use percentiles when testing hypotheses.

Activity 1.3F

1 Which percentile is the same as

a the median **b** the lower quartile

c the upper quartile?

2 **a** Use your cumulative frequency graph for car speeds to find

i the 5th percentile **ii** the 95th percentile.

b Write a sentence that interprets the values you have found in **part a**.

c Estimate which percentile is nearest to 50 mph and write a sentence that interprets your answer.

3 a Find the 10th and 90th percentiles from another cumulative frequency graph you drew when answering **question 2** in **Activity 1.3E.**

b Interpret the values you have found in **part a**.

c Estimate which percentile is nearest to 60 mph and interpret your answer.

4 a On the same axes draw cumulative frequency graphs showing the hours worked by men and women in the UK. (Data originally given in **Activity 1.3D.**)

b Find the median and interquartile range for each dataset.

c Use your answers to **parts a and b** to compare and contrast the usual weekly hours of work of men and women in the UK.

d Compare your answer to **part c** with your answer to **question 2** in **Activity 1.3D.**

	Workers (thousands)	
Hours of work	Men	Women
Less than 6 hours	100	305
6–15 hours	519	1555
16–30 hours	941	3731
31–45 hours	8702	5904
Over 45 hours	5381	1333

Alcohol

The tables show results of a survey carried out to find out about the drinking habits of men and women in England.

Weekly consumption (units of alcohol)	Number of men
0	393
Under 1	450
1–10	2080
11–21	1237
22–35	787
36–50	337
51 and over	337
Total	5621

Weekly consumption (units of alcohol)	Number of women
0	933
Under 1	1266
1–7	2465
8–14	1066
15–25	643
26–35	178
36 and over	110
Total	6661

Opinions vary about what is a 'sensible' amount to drink and what is 'excessive drinking'. The limits shown in the table on theright are given in the *Medical Students Handbook* (MCA 1998).

	Units per week, x	
	Men	Women
Low risk	$x < 21$	$x < 14$
Hazardous	$21 \leqslant x \leqslant 50$	$14 \leqslant x \leqslant 35$
Harmful	$x > 50$	$x > 35$

In December 1995, following a Department of Health review, the government suggested that adults should limit their alcohol consumption to 3–4 units per day for men and 2–3 units per day for women.

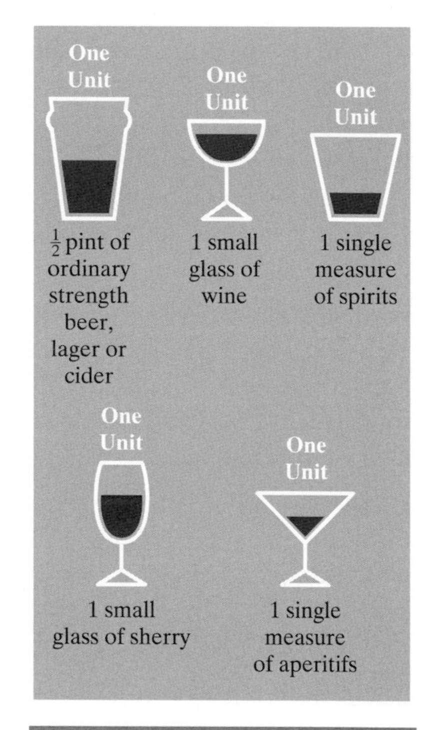

You can calculate the number of units by multiplying the volume of drink in millilitres by the percentage ABV (alcohol by volume) and dividing by 1000.

Activity 1.3G

Write a magazine-style article based on this information. Make sure your article includes appropriate diagrams and measures of location and spread to compare and contrast data. You do not need to use all of the information given – aim to cover 1–2 sides of A4.

Discussion points

Why was the data grouped differently for men and women? What problems will arise if you draw histograms or cumulative frequency graphs to illustrate these data? How will you deal with these problems?

1.4 Revision summary

Collecting data

Primary data is collected directly by you.
Secondary data is collected by someone else and may have been partly processed.
Qualitative data is ***descriptive*** data such as gender, colours, types of employee, subjects studied.
Quantitative data is ***numerical*** data. It can be either discrete or continuous.
Discrete data takes only ***particular***, ***exact values,*** e.g. number of children in a family, shoe sizes.
Continuous data can take ***any value in a continuous range***, e.g. measurements such as height and weight.
Population – all the members of a group being studied, e.g. all cars manufactured by a motor company.
In a **random sample** every member of the population has an ***equal chance*** of being included.
A **stratified sample** is obtained by taking samples from each sub-group (stratum) of a population in proportion to the size of the sub-group in the population. This should give a ***representative*** sample.

Measures of location

Measure of location – average value.

Mode – most common value.

Modal group – group with highest frequency or, if the groups are of different widths, the highest frequency ÷ width.

Median – the middle value in an ordered list. (If there are two middle values, the median is halfway between them.)
The **position** of the median is given by $\frac{1}{2}(n + 1)$ where n is the number of values.

Mean $= \dfrac{\text{sum of values}}{\text{number of values}} = \dfrac{\Sigma x}{n}$

μ denotes the theoretical value of the mean and $\bar{x}$ denotes the mean of a sample.

Mean from a frequency table $\bar{x} = \dfrac{\Sigma xf}{\Sigma f}$

If the data is grouped, x denotes the mid-value and f is the frequency of the group – in this case the value calculated is an estimate of the mean.

Measures of spread

Measure of spread – quantity that measures variability.

Range = Maximum value − Minimum value

Lower quartile (LQ) = $\frac{1}{4}(n + 1)$th value

Upper quartile (UQ) = $\frac{3}{4}(n + 1)$th value

Interquartile range = UQ – LQ

Standard deviation =

$$\sqrt{\frac{\Sigma(x-\bar{x})^2}{n}} \text{ or } \sqrt{\frac{\Sigma x^2}{n} - \bar{x}^2}$$

σ denotes the theoretical standard deviation and s denotes the standard deviation of a sample.

Standard deviation from a frequency table =

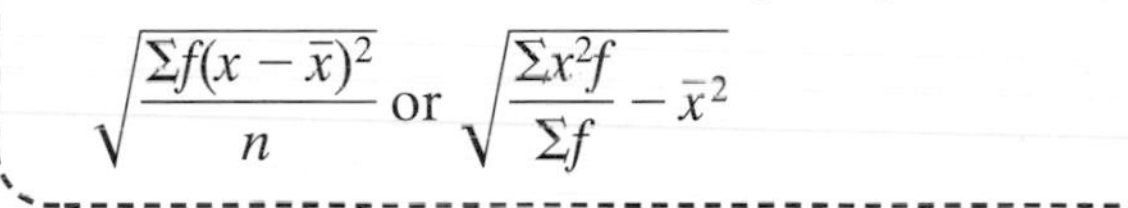

$$\sqrt{\frac{\Sigma f(x-\bar{x})^2}{n}} \text{ or } \sqrt{\frac{\Sigma x^2 f}{\Sigma f} - \bar{x}^2}$$

Statistical charts and graphs

Bar charts are used to illustrate **qualitative** or **discrete** data. Gaps are usually left between each bar and the next.

Comparative bar charts use **adjacent** bars to compare two or more sub-categories.

Component bar charts have bars split into two or more parts representing different sub-categories.

Pie charts show how a total is split into parts. Each sector represents a *proportion* of the total.

Comparative pie charts – the area of each pie chart is proportional to the population it represents. The radii are related by $r_2 = \sqrt{\frac{n_2}{n_1}} \times r_1$ where the pie charts represent populations of n_2 and n_1.

Line graphs can be used to illustrate how a quantity varies with time (time on the horizontal axis).

Histograms are used to illustrate *continuous grouped* data.
There are no gaps between the bars.
Take care when identifying the class boundaries, e.g. in age data, 0–4 means from 0 up to 5 years old.

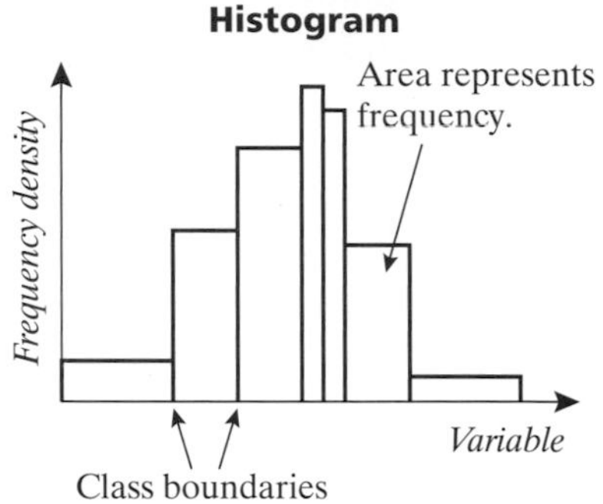

Area represents frequency.

Use frequency density $= \dfrac{\text{frequency}}{\text{class width}}$ on the vertical axis.

Class width = upper class boundary − lower class boundary

Where appropriate, assume open-ended groups are equal in width to an adjacent group; otherwise choose something sensible.

The **cumulative frequency** at a particular value is the number of readings that are *less than* (or less than or equal to) the value. A cumulative frequency table is drawn up by keeping a 'running total' of the frequencies.

Cumulative frequency graph
Points are plotted at the upper boundary of each group. The median, quartiles and percentiles can be found from a cumulative frequency graph. When the total frequency, n, is large, the position of the median, lower quartile and upper quartile can be taken to be $\frac{1}{2}n$, $\frac{1}{4}n$ and $\frac{3}{4}n$ respectively.

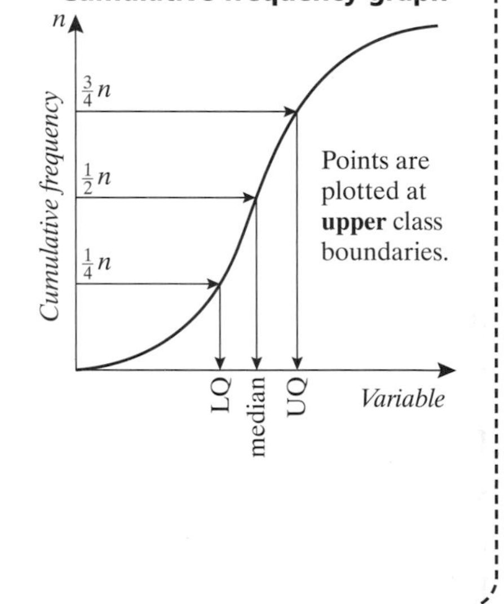

Percentages

Percentage change $= \dfrac{\text{final value} - \text{original value}}{\text{original value}} \times 100$

To *increase* a quantity by, for example, 5%, *multiply* by 1.05; to **reduce** it by 5% *multiply* by 0.95

If x_{now} is 5% more than x_{before}, then $x_{now} = 1.05 \times x_{before}$

so $x_{before} = \dfrac{x_{now}}{1.05}$.

1.5 Preparing for assessment

Your coursework portfolio

In this chapter you have looked at some issues to do with data collection and its representation using a range of diagrams and statistical measures such as mean and standard deviation. At this point you will be beginning to consider how you are going to develop the reports required for your coursework portfolio. You have to write two of these: in one you have to work with data you collect yourself; in the other you should work with secondary data, that is, data you gather from another source.

The first thing to do is to think about what it is you want to investigate. It may be that you have been asked to carry out some statistical work in another part of your studies (in geography, psychology, business studies, biology or another subject). You may be able to use this work to meet the requirements for your *Using and Applying Statistics* coursework portfolio – you are encouraged to do this as it will allow you to demonstrate how you can use statistics to assist you in working on a meaningful problem. On the other hand, you may want to carry out an investigation into something that really interests you so that your work will primarily be for the coursework portfolio for this qualification.

Once you have decided on the area in which you will work, you then need to decide what data you will work with. If the data already exists, that is it is secondary data, it may restrict what you can do. For example, if the data is already grouped you will be unable to go back and find out information about individuals. However, if you are collecting the data yourself, you need to think very carefully about how you will do this. Depending on your chosen investigation you need to consider how you will collect the data. Will you carry out an experiment, take measurements or use a questionnaire?
Whatever method of data collection you are going to use, or what secondary data you will work with, it is important that you consider whether it will allow you to demonstrate as many as possible of the ideas you will meet on this course.

If you are in any doubt about this consult with your teacher as soon as possible.

Data collection by questionnaire

As you have seen, collecting data using a questionnaire has to be done carefully as it is easy to write unclear or ambiguous questions that can be interpreted in different ways by different people. Think carefully about exactly what you want to find out and the conclusions you might want to draw at the end of your piece of work. This should inform your thinking about what questions you need to ask *and* who you need to ask. Of course, you won't be able to involve a large sample in your work, but do make sure that you don't use your questionnaire with only very few people. Don't ask too many questions – people will find your questionnaire time-consuming and you will end up with too much data.

Data collection by experiment or measurement

You may be able to collect data for your investigation by taking measurements or carrying out an experiment. For example, if you were investigating whether or not soaking seeds in water before planting helps them germinate you would carry out an experiment in which you soaked some seeds before planting but not others. You may decide to take into account other factors, for example carrying out the same procedures at two different temperatures. Once you have decided on such matters you can go ahead and carry out your experiment and record the outcomes – your data.

It may be that you do not need to carry out an experiment at all but instead need to record measurements that you take. For example, researchers sometimes try to identify authors by analysing the lengths of words in their writing. You may decide to carry out an investigation like this. Your data collection would therefore involve you in taking measurements based on writing that already exists.

Calculating statistical measures and representing data

You will eventually have some data with which you can work, whether you have collected this yourself or it is secondary data that you have found or been given. At this point you need to think very carefully about which statistical measures are appropriate and you can calculate correctly. You will also want to use diagrams to display the results of your work. Take great care at this stage. It is particularly easy to use spreadsheet software to develop inappropriate and incorrect diagrams (see *Using Technology*).

However, spreadsheet software is very powerful when working with data – you should use it whenever possible, and don't forget that you need to include evidence (print-outs) of having used this to calculate statistical measures and draw diagrams in your coursework portfolio.

Use the work you have done in this chapter to inform the start of the work you need to do for your coursework portfolio.

The Nuffield FSMQ website provides many data sets and web links that may be of use to you.

Practice exam questions

Smoking

Data

In a survey, smokers were asked the age at which they started to smoke.

The table shows the results:

Age	Men	Women	Total
Under 16	1559	1090	2649
16–17	979	892	1871
18–19	544	627	1171
20–24	399	396	795
25 and over	181	264	445

Questions

1 Use 10 as the lower boundary of the first group and 40 as the upper boundary of the last group.

a **i** Calculate the mean and standard deviation of the ages when the **men** started to smoke.

ii Briefly explain why the answers you have given in **part ai** may not be very accurate.

iii The mean age at which the women started to smoke was 18.0 years and the standard deviation was 5.31 years. Use these statistical measures and those you found in **part ai** to compare and contrast the ages when the men and women started to smoke.

b **i** Use the total frequencies in the table to draw histograms to illustrate the ages at which the men and women in the survey began to smoke.

ii Use your histograms to comment on the distribution of ages when the men and women started to smoke.

2 Use 10 as the lower boundary of the first group and 40 as the upper boundary of the last group.

a **i** Draw a cumulative frequency graph to show the ages when the **men** started to smoke.

ii Use your cumulative frequency graph to find the median and interquartile range.

iii The median and interquartile range for the women were 17.2 years and 5.0 years respectively. Use these statistical measures and those you found in **part ai** to compare and contrast the ages when the men and women started to smoke.

b The samples of men and women in the survey were representative samples of male and female smokers in the UK. Briefly explain what this means.

Conceptions

Data

The table gives information about the age at which women conceived children in England and Wales in the years 1991 and 2000.

Notes

1 Conceptions are estimates derived from birth registrations and abortion notifications.

2 Rates give the number of conceptions per thousand women in the age group. Rates for women under 16, under 18, under 20 and 40 and over are based on the population of women aged 13–15, 15–17, 15–19 and 40–44 respectively.

Age	Year	Conceptions (thousands)[1]	Rates[2]
Under 16	1991	7.5	8.9
	2000	8.1	8.3
Under 18	1991	40.1	44.6
	2000	41.3	43.8
Under 20	1991	101.6	64.1
	2000	97.6	62.2
20–24	1991	233.3	120.2
	2000	158.8	104.1
25–29	1991	281.5	135.1
	2000	209.2	116.9
30–34	1991	167.5	90.1
	2000	195.2	95.3
35–39	1991	57.6	34.4
	2000	88.6	42.2
40 and over	1991	12.1	6.6
	2000	17.0	9.3
All ages	1991	853.7	77.7
	2000	766.4	70.6

Question

3 a Calculate the percentage change between 1991 and 2000 in

Resource Sheet 1.5

i the number of conceptions by teenage girls under 16 years of age

ii the total number of conceptions by women of all ages.

b Estimate the number of women in England and Wales in the age group 20–24 in the year 2000.

c The table shows, for 1991, how the conceptions by women of all ages (i.e. women aged 15–44) were distributed between the age groups 15–19, 20–24, 25–29, 30–34, 35–39 and 40–44.

i Complete the table to show the corresponding percentages for the year 2000.

ii Briefly describe the way in which the age of women at conception has changed between 1991 and 2000.

Age	Year	% of conceptions
15–19	1991	11.9
	2000	
20–24	1991	27.3
	2000	
25–29	1991	33.0
	2000	
30–34	1991	19.6
	2000	
35–39	1991	6.7
	2000	
40–44	1991	1.4
	2000	

Consumer prices

Data

The Retail Price Index (RPI) is a measure of consumer prices. The value of the RPI is currently based on a value of 100 taken at the beginning of January 1987.

The most frequent use of the RPI is to calculate the percentage change in prices between one month and the next. The percentage change over a year is usually referred to as 'the rate of inflation'.

The chart shows percentage changes in the RPI for the period from October 1998 to September 2001.

(s.a. means 'seasonally adjusted'.)

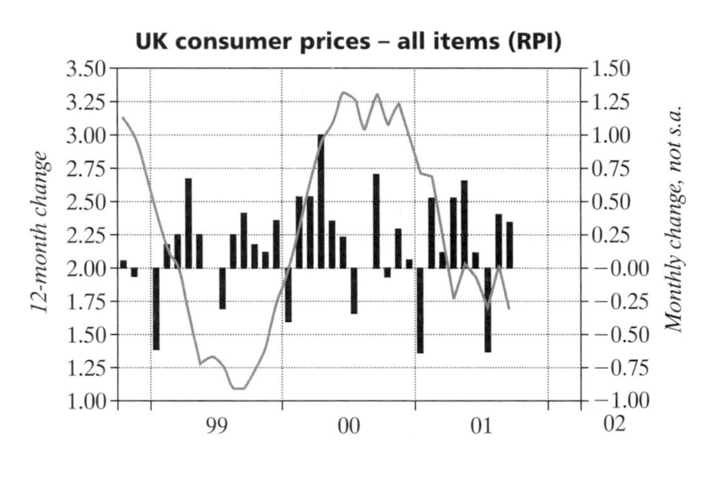

Question

4 a Use the line graph of 12-month changes in the RPI to estimate

i the rate of inflation between the beginning of January 1999 and the beginning of January 2000

ii the percentage change in the RPI between the beginning of January 1999 and the beginning of January 2001.

b i Estimate the monthly change in the RPI in January 2000.

ii If the RPI at the beginning of January 2000 was 167, estimate the RPI at the beginning of February 2000 and at the beginning of March 2000.

iii For which months in the years 1999 and 2000 was there no change in the RPI?

iv What pattern can you identify in the monthly changes in the RPI during the first two months of each year? Suggest a reason for this pattern.

c At the beginning of January 1987, the RPI was taken to be 100 and by the beginning of January 2000 the RPI had increased to 167. If a basket of groceries cost £50 at the beginning of January 1987, what would you expect a similar basket of groceries to have cost at the beginning of January 1987?

2 Correlation and Regression

Contents

The table below shows data for foot length and time to run 100 metres. You can see that as foot length increases, the time taken to run 100 metres decreases. This is shown most effectively in the scatter diagram below.

Foot length (cm)	Time for 100 m (s)
23.4	12.9
25.4	11.9
26.5	11.3
26.9	11.0

What does this suggest? Perhaps if you have a small foot you stand little chance of winning a 100 metre race?

Time for 100 m plotted against foot length

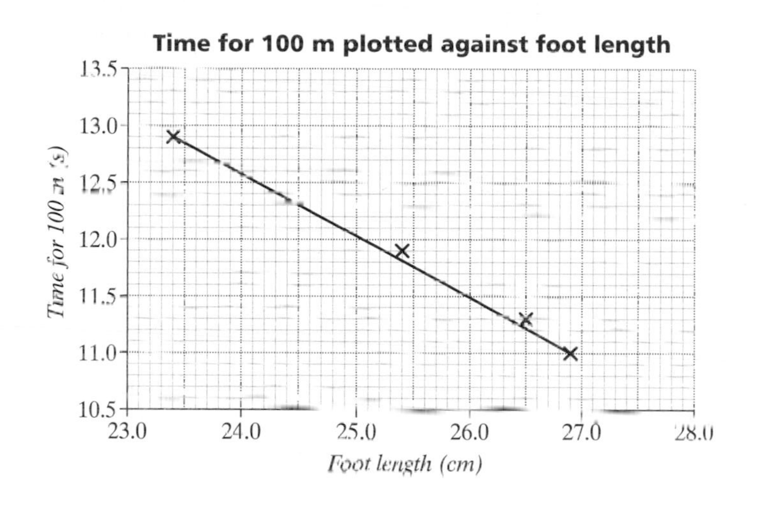

To make sense of the information here you need to know more about the data.

In fact, the data is for boys as they get older as this second version of the table shows. Increasing foot length does not *cause* a decrease in time to run 100 metres. Age is the underlying causal factor!

Age	Foot length (cm)	Time for 100 m (s)
12	23.4	12.9
14	25.4	11.9
16	26.5	11.3
19	26.9	11.0

In this chapter you will learn to explore the relationship between two sets of data using scatter diagrams initially to check to see if there is any link (correlation). You will also learn to calculate measures of this correlation and to find the equation of the line of best fit. Such lines – like the one shown on the diagram above – allow you to predict what will happen in situations for which you have no data. For example, using the graph above you can see that a boy with a foot length of 25 cm might be expected to run 100 m in just over 12 seconds.

Note: when you have two quantities associated with one item (in this case, for each bar you have foot length and time for 100 m) this is known as bivariate data.

2.1 Scatter diagrams

Olympic medals

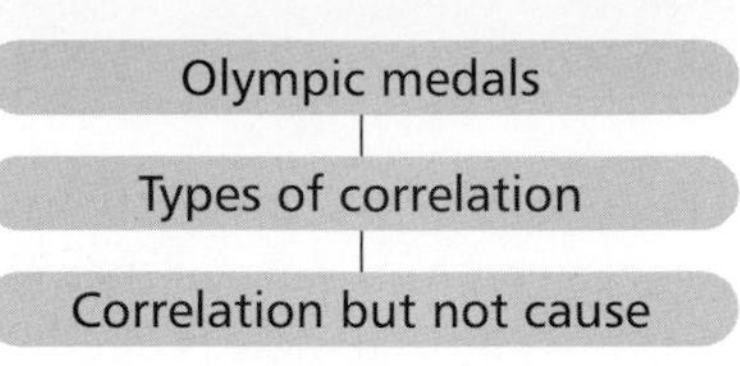

The **scatter diagram** below shows the ten countries that performed best in the 1996 Olympics. The x co-ordinate of the point for each country is its Gross Domestic Product (GDP), a measure of the total wealth of the country. The y co-ordinate is the total score for the medals won in the Olympics – a bronze medal scores one point, a silver medal two points and a gold medal three points.

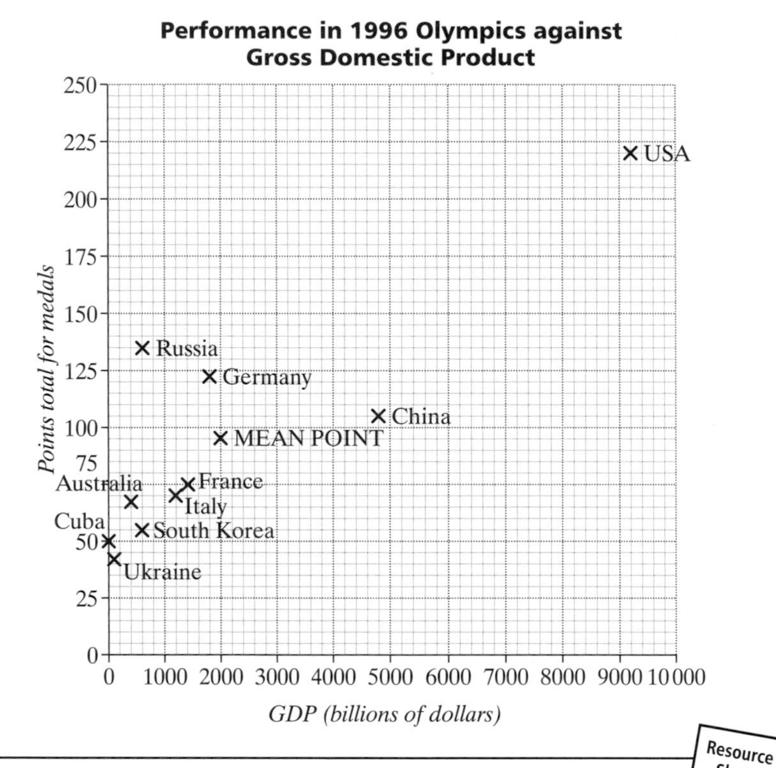

Discussion point

Do you think that there is a causal link between GDP and success in the Olympics (i.e. does having a large GDP lead to Olympic success)? What other factors could affect success?

Activity 2.1A

Resource Sheet 2.1A

1 Describe in words what the scatter diagram shows.

2 Which country or countries stand out from the rest? Why?

3 There is no country in the bottom right of the graph. What can you say about a country that would be in this position?

4 There is no country in the top left of the graph. What can you say about a country that would be in this position?

5 The mean point marked on the graph has the mean GDP as its x co-ordinate and mean medal score as its y co-ordinate. Draw a line of best fit through this point, keeping a balance of countries on each side of the line.

6 Use your line 'of best fit' to estimate what the medal points score could be for a country with a GDP of $4000 billion. Say, with reasons, how accurate you think your estimate is.

Line of best fit

A line of best fit can be

- drawn by eye – passing through the mean point and having a balance of points on either side.
- plotted accurately by using your calculator to find its equation – you will learn how to do this in section 2.3.

Note: It is a common mistake to assume that the line of best fit has to go through the origin. This scatter diagram shows that this is quite wrong; a line through the mean point and through the origin would NOT have a balance of points on each side.

The next scatter diagram shows the medal score for the same 10 countries, but this time shown against the GDP for each person in the country (GDP per capita).

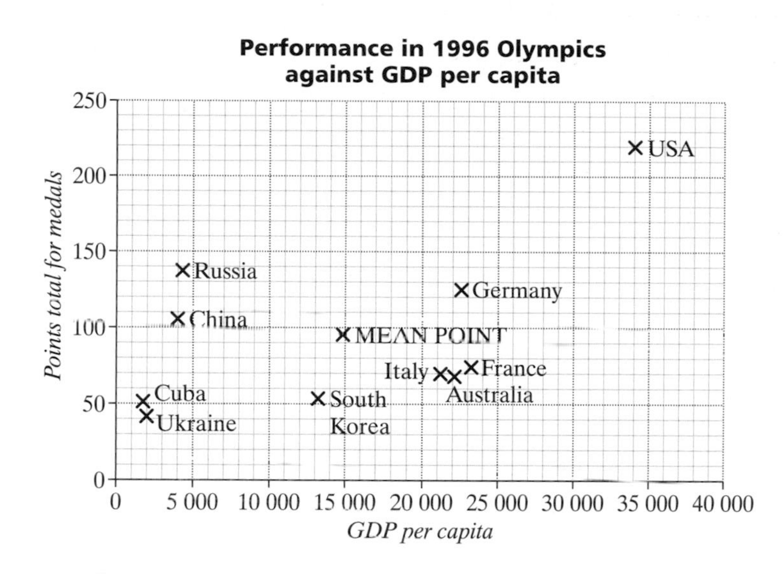

Activity 2.1B

Resource Sheet 2.1B

1. Explain how GDP per capita is calculated from GDP?
2. Which countries appear in very different positions relative to the others in the two scatter graphs? Why?
3. Draw a line 'of best fit' through the mean point, again keeping a balance of countries on each side of the line.

Types of correlation

Both of the scatter diagrams above show **positive correlation**. When one variable is high so is the other; when one variable is low so is the other. The points lie generally on an upward diagonal and there are no points in the top left or bottom right of the diagrams. Higher GDP and higher GDP per capita are both associated with greater success in the Olympics.

As you work through this chapter you will learn a standard way of calculating the strength of correlation. One way of gaining a visual impression is shown below on smaller versions of the two scatter graphs you have met so far. An ellipse enclosing all the points of the scatter graph is drawn; a long narrow ellipse indicates strong correlation – the points lie close to a line of best fit. A 'fatter' ellipse indicates weaker correlation – the points lie further away from a line of best fit. An ellipse that is circular or almost circular indicates no correlation between the two sets of data.

Discussion point

Which of the two scatter diagrams appears to show the stronger correlation?

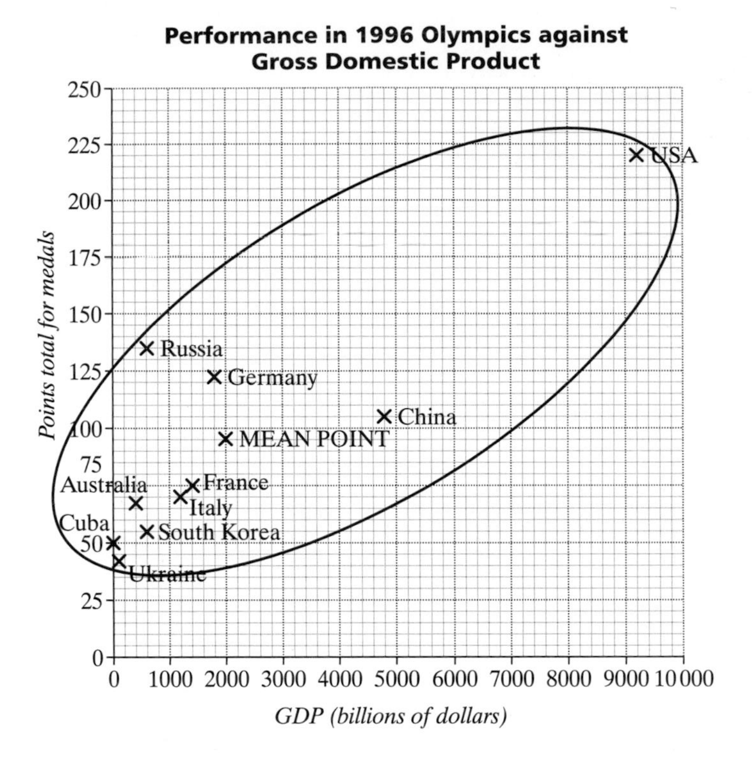

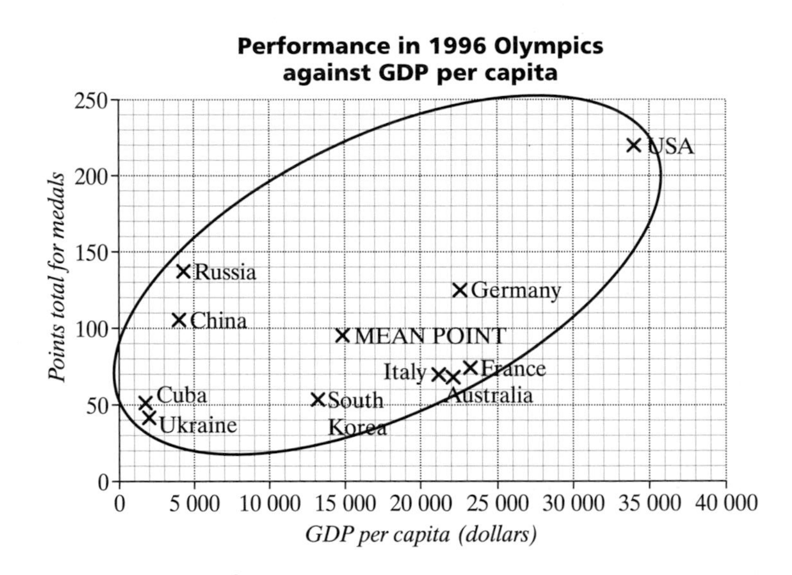

The shape and orientation of the ellipse enclosing all the data points gives an indication of the type and strength of correlation.

Discussion point

Think of some data about people that is not correlated, e.g. date of birth and distance travelled to work.

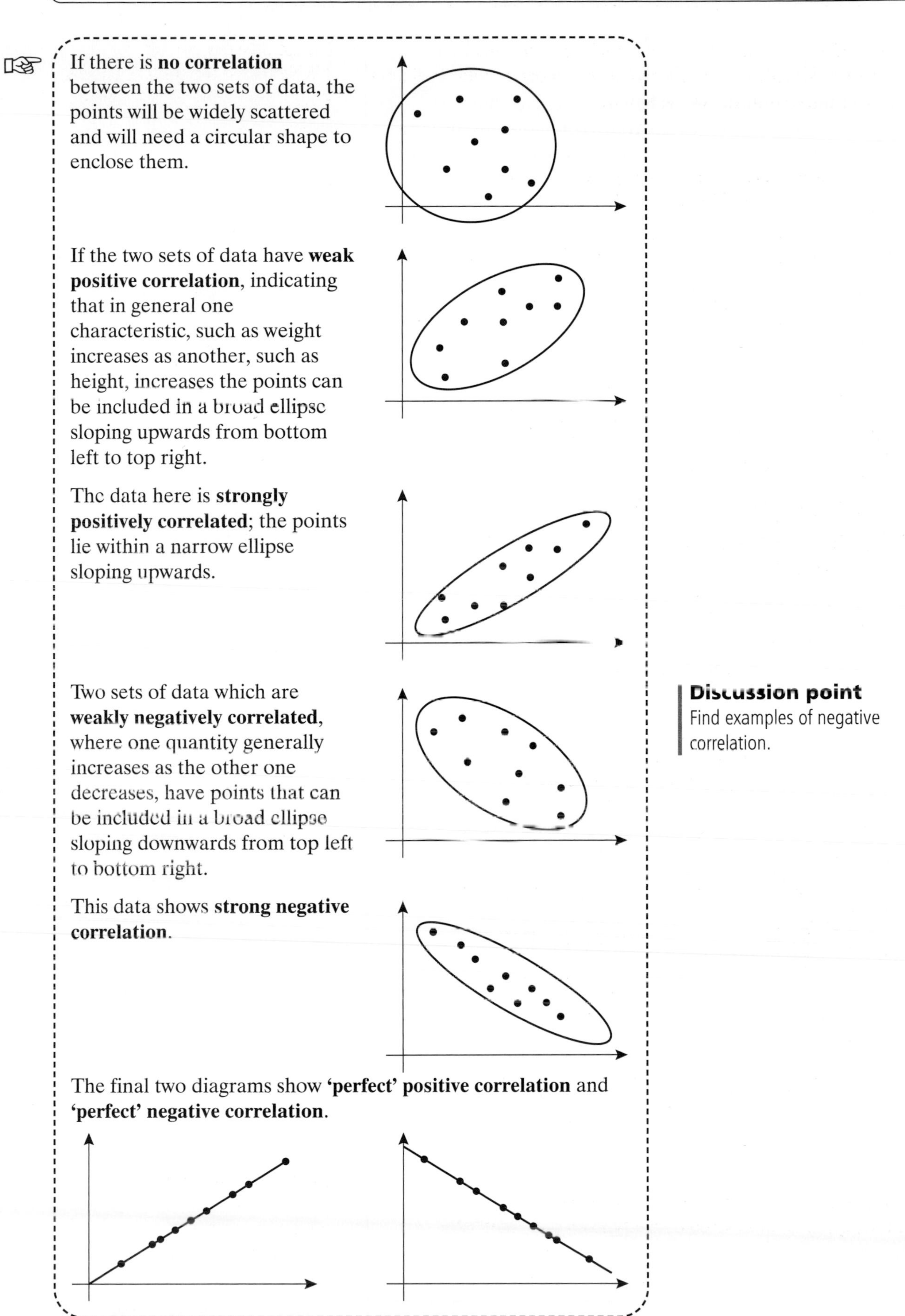

If there is **no correlation** between the two sets of data, the points will be widely scattered and will need a circular shape to enclose them.

If the two sets of data have **weak positive correlation**, indicating that in general one characteristic, such as weight increases as another, such as height, increases the points can be included in a broad ellipse sloping upwards from bottom left to top right.

The data here is **strongly positively correlated**; the points lie within a narrow ellipse sloping upwards.

Two sets of data which are **weakly negatively correlated**, where one quantity generally increases as the other one decreases, have points that can be included in a broad ellipse sloping downwards from top left to bottom right.

This data shows **strong negative correlation**.

The final two diagrams show **'perfect' positive correlation** and **'perfect' negative correlation**.

Discussion point

Find examples of negative correlation.

Activity 2.1C

1 Look at the scatter diagrams below and decide how the correlation is best described: strong correlation, weak correlation, no correlation, positive or negative correlation.

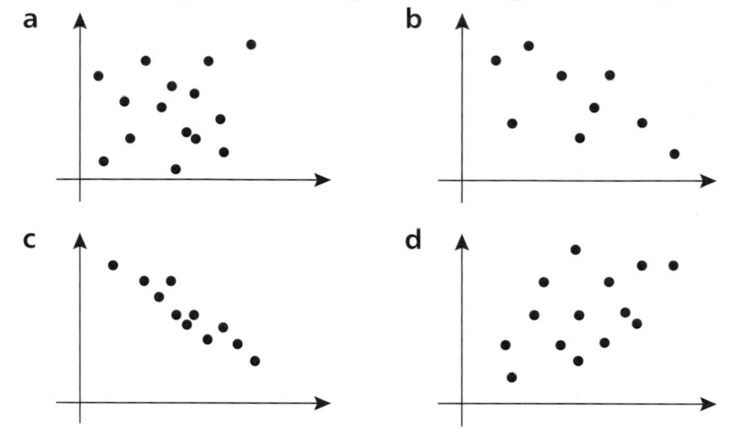

2 For each set of data below, plot one variable against the other in a scatter diagram. Use your scatter diagram to decide whether there is any correlation and whether or not this is positive or negative and whether or not it is weak or strong.

a Average weekly household spending on tobacco and alcohol products by region of Great Britain.

Region	Alcohol (£)	Tobacco (£)
North	£6.47	£4.03
Yorkshire	£6.13	£3.76
Northeast	£6.19	£3.77
East Midlands	£4.89	£3.34
West Midlands	£5.63	£3.47
East Anglia	£4.52	£2.92
Southeast	£5.89	£3.20
Southwest	£4.79	£2.71
Wales	£5.27	£3.53
Scotland	£6.08	£4.51
Northern Ireland	£4.02	£4.56

b Height and age of children in an Egyptian village.

Age (months)	Height (cm)
18	76.1
19	77
20	78.1
21	78.2
22	78.8
23	79.7
24	79.9
25	81.1
26	81.2
27	81.8
28	82.8
29	83.5

Note: You can use your graphic calculator or spreadsheet software on a computer to draw these scatter diagrams.

Discussion point
Does it matter which variable you plot on the horizontal axis and which you plot on the vertical axis? Investigate this by plotting the data for one or more parts in both ways.

Discussion point
Discuss reasons for correlation type.

c Consumption of butter and margarine (in grams per person per week) for people in Great Britain during the period 1971–1981.

Year	Butter	Margarine
1971	153	87
1972	135	99
1973	148	85
1974	160	74
1975	160	74
1976	147	86
1977	133	98
1978	129	100
1979	126	103
1980	115	108
1981	105	116

Correlation but not cause

As shown in the introduction, it is possible for data to be linked but for one factor not to be the cause of the other. For example, the scatter diagram below shows the marks that 15 students were given on two questions on a GCSE mathematics paper. The marks are weakly positively correlated (look at the shape of the ellipse). This means that, generally, a student who did well on question 1 would be likely to do fairly well on question 2 also, but positive correlation does *not* mean that there is a **causal link**, i.e. it does not mean that good performance on the first question *causes* good performance on the second. A better explanation is that a student's ability to do well in mathematics exams causes good performance on both questions.

Note: You can identify only 11 points on this graph although it is for 15 students. This is because some of the students got exactly the same results as each other (see data on page 48).

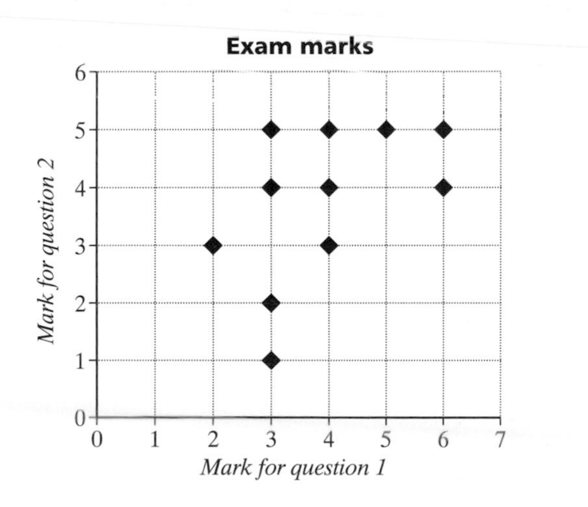

Discussion points

Why do the data points appear in 'lines' across the paper and up and down?

The marks in the two questions are not very closely related. What could be the reasons for this?

Activity 2.1D

1 The table below shows data for six schools in one town for a particular year. It shows the percentage of pupils in each school who claim free school meals and the percentage of pupils in Year 11 who at the end of the year gained five or more higher grade GCSE passes.

	Percentage of pupils claiming free school meals	Percentage of pupils gaining five or more GCSEs at grades A–C
Oakwood	33.3	24.8
St Johns	5.5	58.8
Bridgemount	22.1	42.3
Westford	22.4	33.8
Highfield	30.7	30.8
All Saints	23.5	38.0

a Draw a scatter diagram of the data for the six schools.

b What sort of correlation does your scatter diagram show?

c Interpret what your graph tells you in terms of GCSE results and the percentages of pupils receiving free school meals.

d Calculate the mean values of each of the two sets of data. Plot this point on your scatter diagram and draw a line of best fit through this point.

e Would the Director of education for this town be justified in saying, 'If we make all the pupils pay for their school meals GCSE results will improve'? Explain your answer.

2.2 Calculating correlation coefficients

Alcohol concern

Alcohol concern
Exam marks
Interpreting correlation coefficients

The table shows data taken from *State of the Nation 2002* released by Alcohol Concern. It gives data for each region of England about the percentage of men and women whose weekly consumption of alcohol is above the recommended safe levels of 21 units for men and 14 units for women.

Region	Men x	Women y
North West	45%	27%
North East	44%	26%
Yorkshire & Humber	42%	23%
Merseyside	46%	28%
East Midlands	43%	23%
West Midlands	35%	19%
East of England	31%	20%
South East	39%	23%
South West	35%	21%
London	31%	19%

The data is plotted as a scatter diagram below.

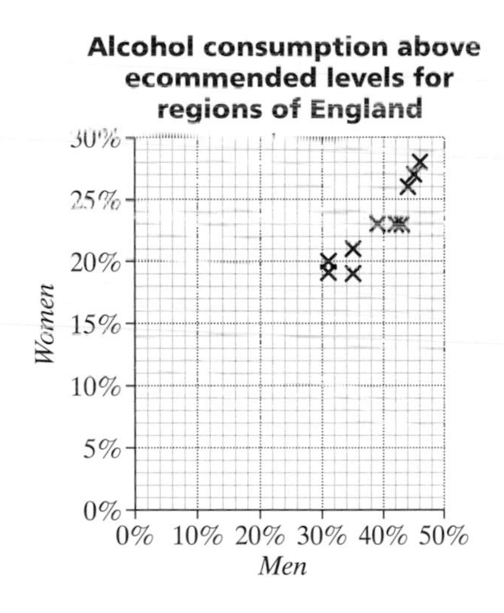

In the following activity you will find the **correlation coefficient** for the two sets of data by working through the calculations. This will give you an overview of how such correlation coefficients are found – you do not usually need to work at this level of detail but you may find it informative.

Activity 2.2A

1 Use the 'Alcohol consumption above recommended levels for regions of England' data, to find the mean percentage for men, $\bar{x}$, and the mean percentage for women data $\bar{y}$.

2 The table is repeated below with percentages converted to decimal values for ease of calculation. Complete a copy of this table finding values for $x - \bar{x}$ and $y - \bar{y}$ for each region (some of the values have been calculated for you).

Region	Men	Women		
	x	y	$x - \bar{x}$	$y - \bar{y}$
North West	0.45	0.27	0.059	
North East	0.44	0.26	0.049	
Yorkshire & Humber	0.42	0.23	0.029	
Merseyside	0.46	0.28	0.069	
East Midlands	0.43	0.23	0.039	
West Midlands	0.35	0.19	−0.041	
East of England	0.31	0.20		
South East	0.39	0.23		
South West	0.35	0.21		
London	0.31	0.19		

Discussion points

What can you deduce from this table of data and the scatter diagram?

What further information do you need to know about how data was gathered?

If the value of $(x - \bar{x})$ is positive it means that the men in the region are more likely to exceed the recommended safe drink limit than average.

3 Interpret, in terms of the original data, what each of the following means:

a a negative value of $x - \bar{x}$

b a positive value of $y - \bar{y}$

c a negative value of $y - \bar{y}$.

- If there is strong correlation between the two sets of data then, for a particular region, if $x - \bar{x}$ is positive it is likely that $y - \bar{y}$ will also be positive; and if $x - \bar{x}$ is negative it is likely that $y - \bar{y}$ will also be negative.

- It is therefore likely that the product $(x - \bar{x})(y - \bar{y})$ will be positive.

4 Confirm the above statement by finding the product $(x - \bar{x})(y - \bar{y})$ for the data above and recording this in an additional column of your copy of the table.

Do not be concerned about the units of the values you are calculating.

Region	Men	Women			
	x	y	$x - \bar{x}$	$y - \bar{y}$	$(x - \bar{x})(y - \bar{y})$
North West	0.45	0.27	0.06		
North East	0.44	0.26	0.05		
Yorkshire & Humber	0.42	0.23	0.03		
Merseyside	0.46	0.28	0.07		
East Midlands	0.43	0.23	0.04		
West Midlands	0.35	0.19	−0.04		
East of England	0.31	0.20			
South East	0.39	0.23			
South West	0.35	0.21			
London	0.31	0.19			

5 The sum of the values you have just calculated is denoted by S_{xy}. Calculate this value.

It was proposed that a measure of the co-relationship between two variables could be based on this measure, S_{xy}.

The sum of the squares of $(x - \bar{x})$ gives the value S_{xx}.

The sum of the squares of $(y - \bar{y})$ gives the value S_{yy}.

The measure – the **product moment or Pearson's correlation coefficient** is given by:

$$r = \frac{S_{xy}}{\sqrt{S_{xx}S_{yy}}}.$$

6 Extend your copy of the table to add columns in which you calculate $(x - \bar{x})^2$ and $(y - \bar{y})^2$.

Use the values you have calculated in your copy of the table to show that the product moment correlation coefficient for the data is 0.92.

Exam marks

In **Section 2.1** you looked at a scatter diagram of the marks that 15 students gained in two GCSE mathematics questions. The scatter diagram showed weak positive correlation. This means that in general, students doing well on one question were likely to do quite well on the other.

In this activity you will work through calculations that will show another method of finding the correlation coefficient. On the next page is a table of the students' marks, with some extra columns added which you can use to carry out calculations.

A	B	C	D	E	F
Student	Mark in first question (X)	Mark in first question (Y)	Product of marks in first and second questions (XY)	Square of mark in first question (X^2)	Square of mark in second question (Y^2)
1	6	5			
2	4	5			
3	4	5			
4	5	5			
5	4	5			
6	4	4			
7	3	4			
8	6	4			
9	3	5			
10	4	4			
11	3	1			
12	4	3			
13	4	5			
14	3	2			
15	2	3			
Totals:	$\Sigma X=$	$\Sigma Y=$	$\Sigma XY=$	$\Sigma X^2=$	$\Sigma Y^2=$

Note: In this table you are calculating different values to those you found in Activity 2.2A to find the correlation coefficient. You will eventually use a different method to calculate the correlation coefficient.

Activity 2.2B

1. For each student find the products of their marks for both questions and the square of the mark they obtained in each question to complete columns D, E and F.
2. Find the sums of
   - **a** the marks for each question (columns B and C), i.e. ΣX and ΣY
   - **b** the products of the marks (column D), i.e. ΣXY
   - **c** the squares of the marks for each question (columns E and F), i.e. ΣX^2 and ΣY^2 (see the bottom row of the table).
3. Finally, substitute these values into the following formula to calculate the product moment correlation coefficient, r:

$$r = \frac{N\Sigma XY - \Sigma X \Sigma Y}{\sqrt{(N\Sigma X^2 - (\Sigma X)^2)(N\Sigma Y^2 - (\Sigma Y)^2)}}$$

where N (=15) is the number of pairs of data.
Show that $r = 0.47$.

Checkpoint
Check your calculations by working them out in a different way or by comparing them with someone else's answers.

Note: This is a different formula to that used in the previous activity. You might like to check that both formulae give the same value for r, by recalculating r for each set of data using the alternative method.

Interpreting correlation coefficients

All correlation coefficients lie between –1 and +1 (inclusive).

A coefficient of **+1** indicates **perfect positive correlation**; all the points lie exactly on a straight line with positive gradient.

A coefficient of **–1** indicates **perfect negative correlation**; all the points lie on a straight line with negative gradient.

Here is a rough and ready guide you can use to interpret values of Pearson's product moment correlation coefficient.

Fortunately, graphic calculators allow you to calculate the correlation coefficients by entering the data into a table and then pressing, at most, a few keys.

Below are the steps you need to carry out with a graphic calculator to find the correlation coefficient for the exam mark data.

Sketch	Value of correlation coefficient, r	
	$r = -1$	Perfect negative correlation
	$-1 < r \leqslant -0.5$	Strong negative correlation
	$-0.5 < r < 0$	Weak negative correlation
	$r = 0$	No correlation
	$0 < r \leqslant 0.5$	Weak positive correlation
	$0.5 < r < 1$	Strong positive correlation
	$r = 1$	Perfect positive correlation

In this course, you only need to consider linear (straight line) correlation, though other types do exist.

Enter the data in a table.

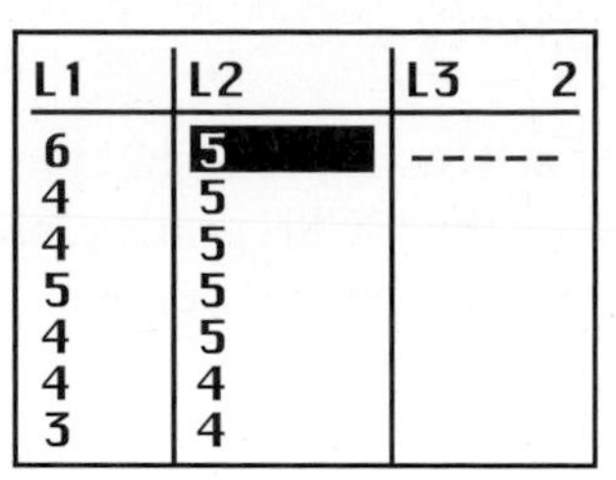

L1	L2	L3 2
6	5	-----
4	5	
4	5	
5	5	
4	5	
4	4	
3	4	

Use the functions that give the correlation coefficient.

```
LinReg
  y=ax+b
  a=.531496063
  b=1.909448819
  r²=.2174302076
  r=.4662941213
```

Plot a scatter diagram (here the line of best fit is shown – you will learn more about this in the next section).

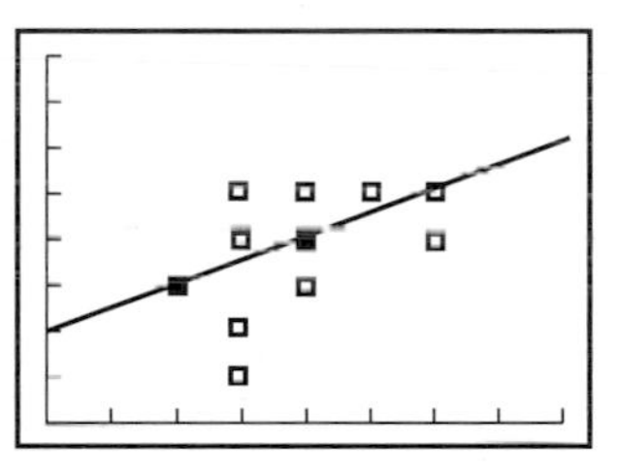

Find out what your calculator will do to help you work out correlation coefficients and practise so that you can use it quickly and accurately.

Note: If you are asked to interpret the correlation between two sets of data, you should indicate the type of correlation and the strength in terms of the original situation. For example, in Activity 2.2A, there is a strong positive correlation between the levels of consumption of alcohol above recommended levels for men and women in the regions of England.

Activity 2.2C

1 a Use a spreadsheet to find the correlation coefficient for the GCSE grades/free school meals data. The table of data is repeated below.

	Percentage of pupils claiming free school meals	Percentage of pupils gaining five or more GCSEs at grades A–C
Oakwood	33.3	24.75
St Johns	5.5	58.75
Bridgemount	22.1	42.25
Westford	22.4	33.75
Highfield	30.7	30.75
All Saints	23.5	38.0

b What does the correlation coefficient tell you about the data?

2 Using the following data, investigate to see if there is any correlation between length and wingspan and length and weight of eagles by finding the appropriate correlation coefficients.

Name	Length (cm)	Wingspan (cm)	Weight (g)
Phillippine Eagle	102	200	8000
Spanish Imperial Eagle	85	220	3500
Bald Eagle	96	244	6300
Wedge-tailed Eagle	104	253	5300
Martial Eagle	86	212	6200
Steller's Sea Eagle	94	215	9000
Harpy Eagle	105	200	9000

3 The following data are for men in 25 occupational groups in England. The smoking index is the ratio of the average number of cigarettes smoked per day by men in the particular occupational group to the average number of cigarettes smoked per day by all men. The mortality index is the ratio of the rate of deaths from lung cancer among men in the particular occupational group to the rate of deaths from lung cancer among all men.

Find out whether there is any link between smoking and mortality rate by calculating the correlation coefficient for the data.

Occupational group	Smoking	Mortality
Farmers, foresters and fishermen	77	84
Miners and quarrymen	137	116
Gas, coke and chemical makers	117	123
Glass and ceramic makers	94	128
Furnace, forge, foundry and rolling mill workers	116	155
Electrical and electronics workers	102	101
Engineering and allied trades	111	118

Discussion point
Before you do the calculation, think about what sort of answer you will get – will it be positive or negative? Near to zero or near to +1 or −1?

Pearson's correlation coefficient can be calculated automatically on some spreadsheets using the function 'Pearson'. You can also use the spreadsheet semi-automatically, by calculating the column totals then feeding them into one of the formulae given above. You should, of course, get the same result either way!

Nuffield Resource Skills activities 'Casio calculators' 'Correlation'

Occupational Group	Smoking	Mortality
Woodworkers	93	113
Leather workers	88	104
Textile workers	102	88
Clothing workers	91	104
Food, drink and tobacco workers	104	129
Paper and printing workers	107	86
Makers of other products	112	96
Construction workers	113	144
Painters and decorators	110	139
Drivers of stationary engines, cranes, etc.	125	113
Labourers not included elsewhere	133	146
Transport and communications workers	115	128
Warehousemen, storekeepers, packers and bottlers	105	115
Clerical workers	87	79
Sales workers	91	85
Service, sport and recreation workers	100	120
Administrators and managers	76	60
Professsionals, technical workers and artists	66	51

Write a sentence or two in which you interpret what you find.

4 Calculate the correlation coefficient for points that lie on $y = 2x + 3$. Use x values from -3 to 3 and corresponding y values calculated using the straight-line equation. Show that the answer is 1.

5 Explain why data of people's height in feet and inches and their height in metres would have a correlation coefficient of 1.

6 In the first section of this chapter you looked at scatter diagrams showing data about Gross Domestic Product (GDP) and Olympic success and compared the data for GDP and GDP per capita. Use correlation coefficients to find out whether the Olympic medals score is correlated more strongly with GDP or with GDP per capita.
Write a sentence or two to explain what you find. Your explanation should refer clearly to the real situation.

7 The table gives details of the masses of various mammals together with the masses of their brains.

a Plot a scatter diagram of the data.

b Use your graphic calculator to calculate the correlation coefficient for the data.

c Write a brief sentence to explain what the diagram and the correlation coefficient tell you about how the data are related.

d Calculate

i the mean body mass, $\bar{x}$ **ii** the mean brain mass, $\bar{y}$.

e **i** On your scatter diagram plot by eye a line of best fit.

ii Use this line to predict y when $x = 75$.

iii Use this line to predict y when $x = 0$. Comment on this answer.

Practice Sheet Product – Moment Correlation

Name	Body mass (kilograms	Brain mass (grams)
	x	y
Baboon	11	180
Chimpanzee	52	440
Cow	465	423
Deer (roe)	15	98
Donkey	187	419
Giraffe	529	680
Goat	28	115
Gorilla	207	406
Horse	521	655
Jaguar	100	157
Kangaroo	35	56
Pig	192	180
Seal (grey)	85	325
Sheep	56	175
Wolf (grey)	36	120

2.3 Regression

Lines of best fit

Lines of best fit

Regression lines

Regression of y on x and x on y

Here is the scatter diagram for the GCSE results against free school meals data that you used earlier. The mean point (the point with co-ordinates that are the mean of the free school meals percentages and the mean of the GCSE results percentages i.e. (22.9, 38.1)) has been added to the graph.

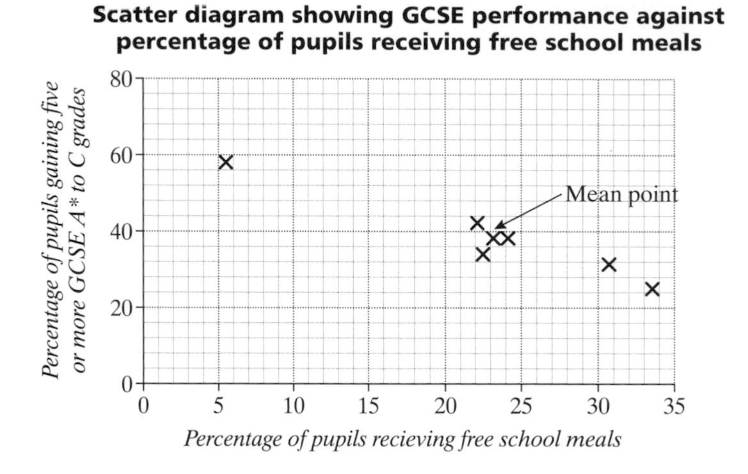

There is strong negative correlation between these two sets of data, as you know from the correlation coefficient that you calculated in the last section. It's also obvious from the fact that the points follow a linear formation quite closely; an ellipse enclosing the points would be long and narrow.

Discussion point

Why do you think that schools with a high proportion of pupils receiving free school meals are likely to have low GCSE scores?

The **line of best fit** goes through the mean point and is angled so that there is a balance of points on each side of the line. You could use the line of best fit to estimate what percentage of pupils would gain five or more GCSE A* to C grades if you know what percentage of pupils receive free school meals in a school. Using the line to predict such results does not necessarily assume that there is a *causal* connection between free school meals and GCSE results, just that the two percentages may be linked.

Activity 2.3A

Resource Sheet 2.3A

1 On a copy of this scatter diagram, draw a line of best fit.

2 Use the line of best fit to estimate what percentage of pupils would get five or more GCSE A* to C grades in a school where 15% of pupils receive free school meals.

3 Find an equation for your line in the form $g = as + b$, where g is the percentage of pupils with five or more high grade GCSEs, s is the percentage of pupils with free school meals and a and b are constant values.

4 Use your equation to calculate an estimate of the percentage of pupils that would get five or more GCSE A* to C grades in a school where 15% of pupils receive free school meals. Compare your result with your answer to **question 2**.

Remember the general form of the equation of a straight line is $y = mx + c$, where m is the gradient and the line crosses the y axis at $(0, c)$.

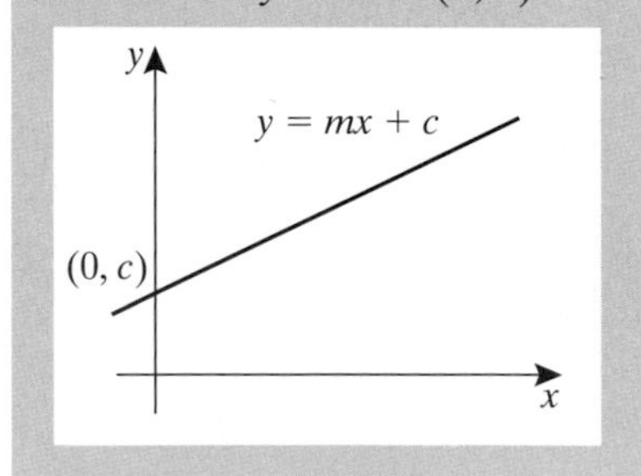

Discussion point

How accurate do you think your estimate is?

Regression lines

You can use the lines of best fit you have just drawn to find some useful estimates, but of course your line of best fit may not be the same as someone else's, so it is helpful to have a systematic method that always gives the same result. One procedure commonly used is the **'method of least squares'**.

On the scatter graph of Olympic results against GDP per capita shown below, one possible line of best fit has been drawn through the mean point. Lines parallel to the *y* axis indicate the distance of each point from the line of best fit. The more closely the line fits the points, the better it is, so the aim is to make the sum of the distances as small as possible.

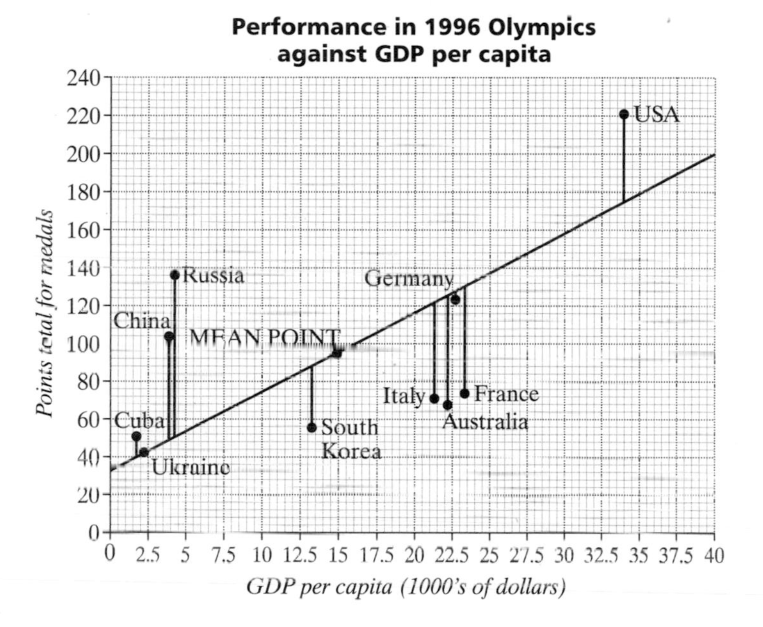

Some of the points are below the line and some above. If the distances from the line are calculated by subtracting *y* co-ordinates, some will be negative and some will be positive, so may cancel out when added. So the square of each distance, which will be positive whether the distance is positive or negative, is used instead. The equation of the line is found by making the total of the squares of the distances as small as possible, hence the name 'method of least squares'.

Fortunately, you do not have to know the method for minimising the sum of the squares, but you do have to know how to use the resulting formula. The equation of the line of best fit according to the method of least squares is $y = ax + b$, with gradient a and y intercept b. A line of best fit determined in this way is called a **regression line**.

In the next activity you will find values of a and b using formulae. Follow this procedure and also find out how to use your calculator to find a and b more easily.

Note: When calculating standard deviation, the squares of deviations from the mean are used for the same reason. You have also just seen in **Section 2.2** that when calculating correlation coefficients the squares of values are used.

You can find the values of a and b in the regression line $y = ax + b$ using:

$$a = \frac{N\Sigma XY - \Sigma X\Sigma Y}{N\Sigma X^2 - (\Sigma X)^2}$$

$$b = \frac{\Sigma X^2\Sigma Y - \Sigma X\Sigma XY}{N\Sigma X^2 - (\Sigma X)^2}$$

Both formulae contain some elements familiar from the last section.

Activity 2.3B

Resource Sheet 2.3B

1 Use a spreadsheet to calculate the values of a and b for the equation of the regression line.

	A	B	C	D	E
1	**Country**	**GDP per capita in dollars (X)**	**Medal points in 1996 Olympics (Y)**	***XY***	***X* squared**
2	United States	33900	221		
3	Russia	4200	136		
4	Germany	22700	123		
5	China	3800	104		
6	France	23300	74		
7	Italy	21400	71		
8	Australia	22200	68		
9	South Korea	13300	56		
10	Cuba	1700	51		
11	Ukraine	2200	43		

a Calculate the products of each pair of X and Y values and the squares of the X values in columns D and E of the spreadsheet.

b Find

- **i** ΣX in cell B12
- **ii** ΣY in cell C12
- **iii** ΣXY in cell D12
- **iv** ΣX^2 in cell E12.

c Substitute these values in the formulae to find a and b.

$$a = \frac{N\Sigma XY - \Sigma X\Sigma Y}{N\Sigma X^2 - (\Sigma X)^2}$$

$$b = \frac{\Sigma X^2\Sigma Y - \Sigma X\Sigma XY}{N\Sigma X^2 - (\Sigma X)^2}$$

Hence find the equation of the regression line.

d Use the spreadsheet to plot a scatter diagram of the data together with the regression line.

e How well does the regression line fit the data?

f What meaning can you give to the point where the regression line cuts the y axis?

g What meaning can you give to the gradient of the regression line?

2 Use your calculator to find the equation of the regression line for the free school meals data (page 50) and compare it with the equation you found for the line of best fit.

Using a graphic calculator to find the regression line

L1	L2	L3 1
33900	221	-----
4200	136	
22700	123	
3800	104	
23300	74	
21400	71	
22200	68	

Enter data into a table.

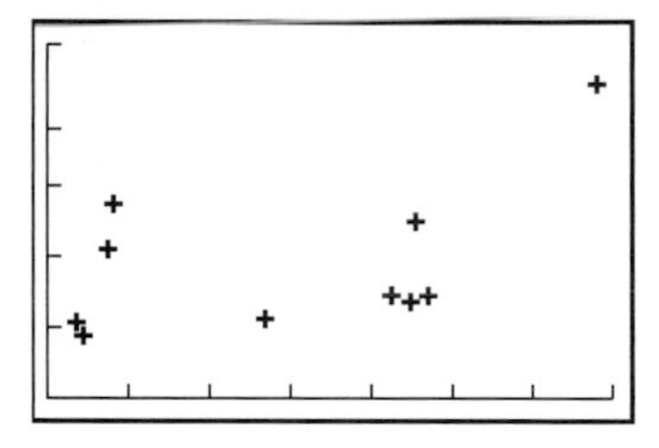

A scatter diagram of the data.

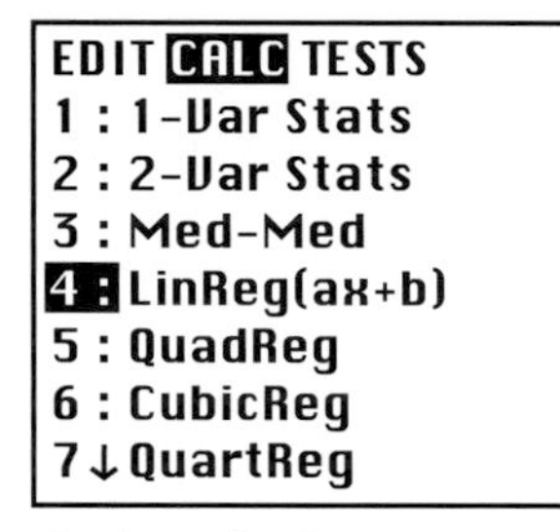

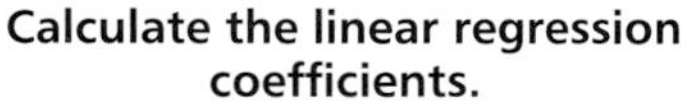

Calculate the linear regression coefficients.

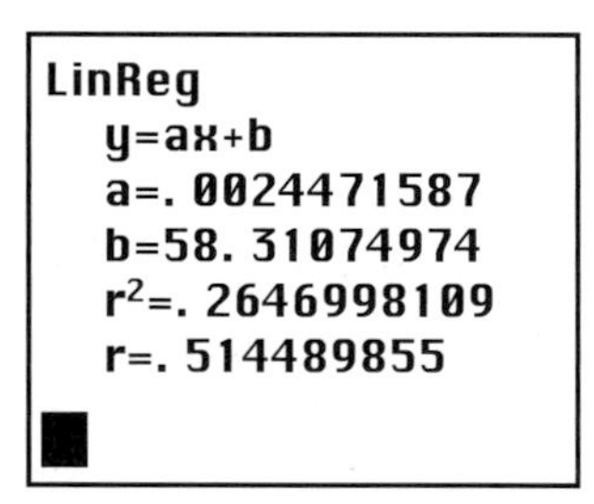

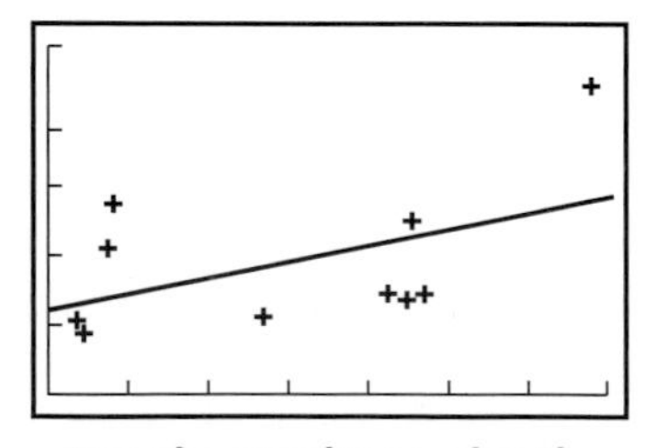

Use the results to plot the regression line together with the data.

3 The table shows the temperature and the relative humidity at one place at regular intervals during one day. (The relative humidity is the ratio of the amount of water vapour in the air to the maximum amount that the air could hold.) This data is plotted in the scatter diagram below.

Temperature (degrees Fahrenheit)	Relative humidity (per cent)
x	y
66	52
68	52
68	53
70	45
72	42
74	33
78	32
81	28
79	30
78	31
77	31
75	32

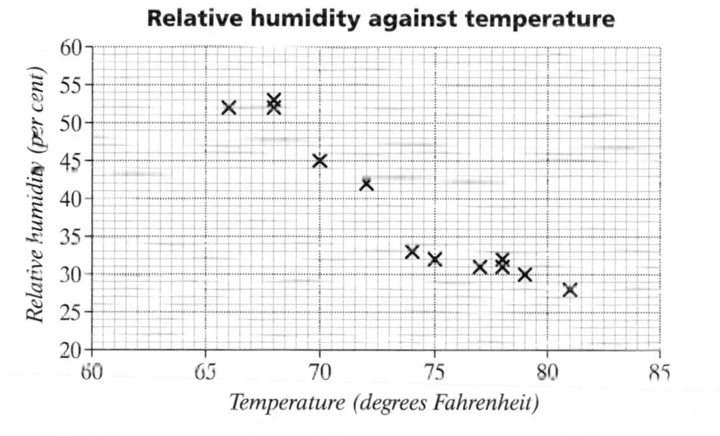

a Find the mean value of the temperature and the mean value of the humidity. Mark the mean point on a copy of the scatter diagram.

b **i** Use your calculator to find the equation of the line of regression, $y = ax + b$.

ii Draw this line of regression on your copy of the scatter diagram.

c Use the regression line to estimate

i the relative humidity for a temperature of 74 degrees Fahrenheit

ii the relative humidity for a temperature of 85 degrees Fahrenheit.

iii State which of your two estimates you expect to be the more accurate and explain why.

d Explain why the regression line you have drawn may not give a good estimate of the temperature when the relative humidity is 35%.

4 The table shows, for various levels of education in the United States, median annual earnings and unemployment rate.

a Write a sentence or two interpreting how median annual earnings and percentage unemployment rate vary with increasing educational achievement.

b **i** Draw a scatter diagram to display the quantitative data in the table.

ii Comment on the relationship between the two sets of data.

c **i** Calculate the mean point of the data and mark it on a copy of the scatter diagram.

ii Use your calculator to find the equation of linear regression, $y = ax + b$, of unemployment rate on median annual earnings for the data.

iii Draw the line of regression on your copy of the scatter diagram.

d Give reasons why it would not be appropriate to use this regression line to estimate the median annual earnings for an educational group with an unemployment rate of 3%.

Highest educational qualification	Median annual earnings, *x* (dollars)	Percentage unemployment rate, *y*
Professional degree	71 000	1.3
Doctorate	62 400	1.4
Master's degree	50 000	1.6
Bachelor's degree	40 100	1.9
Associate degree	31 700	2.5
College but no degreee	30 400	3.2
High school diploma	26 000	4.0
No high school diploma	19 700	7.1

Regression of *y* on *x* and *x* on *y*

The diagram below shows some bivariate data, the lower arm lengths, *x*, and lower leg lengths, *y*, of a group of boys. The straight line is the line of regression of *y* on *x* – the total of the squares of the distances of the data points above and below the regression line, shown as solid vertical lines, has been made as small as possible. For this data the equation of the regression line is $y = 0.818x + 16.68$.

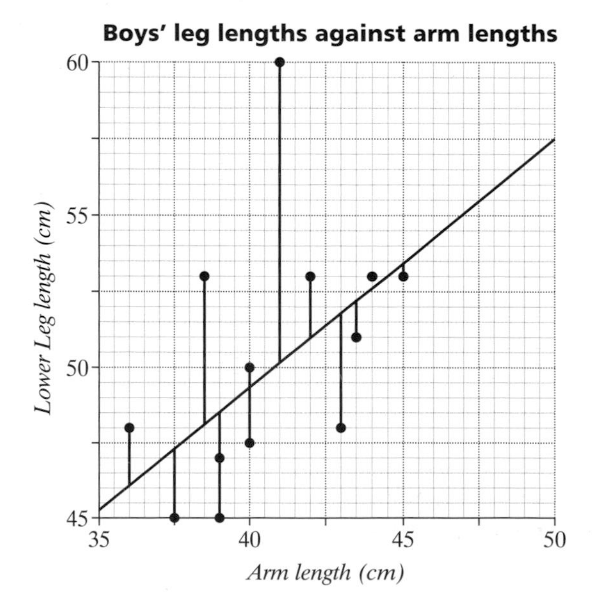

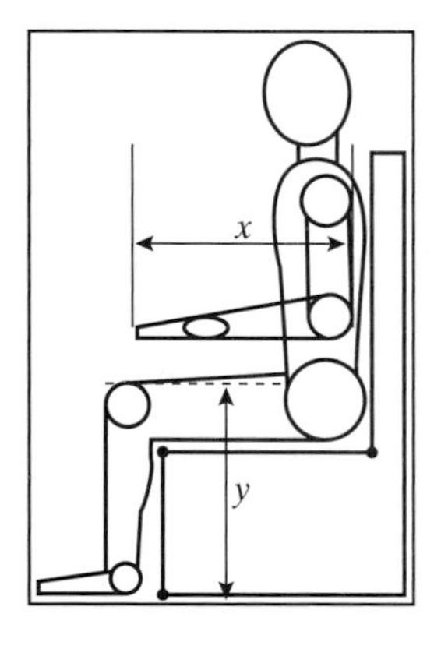

Activity 2.3C

1 Use your calculator to check that $y = 0.818x + 16.68$ is the correct equation for the regression line. You will need to use the raw data given in the table.

2 Use the regression line to estimate the leg length for a boy whose arm length is

a 41 cm b 50 cm.

Comment on the likely accuracy of your answers.

3 One of the boys is an obvious 'outlier'.

a Use a spreadsheet or calculator to find the equation of the regression line for the data with the outlier removed.

b Draw this new line on the graph and use it to repeat the estimates in **question 1**.

c Comment on the effect on your estimates of removing the outlier.

	A	B	C	D
1	Arm and leg data			
2				
3	Boys		Girls	
4	Arm (cm)	Leg (cm)	Arm (cm)	Leg (cm)
5	36	48	39	43
6	37	44	39	44
7	37.5	45	40	48
8	38	53	40	49
9	39	45	41	45
10	39	47	41	48
11	39	47	41	49
12	40	47.5	41	50
13	40	50	42	43
14	41	60	42	46
15	42	53	42	49
16	43	48	42	50
17	43.5	51	43	45
18	44	53	43	49
19	45	53	43.5	45

The regression line of y on x is appropriate to estimate y values (leg lengths) from x values (arm lengths). If you want to work the other way, that is to predict from leg lengths to arm lengths you need to find a new regression line on the original graph so that the total of the squares of the horizontal distances of the data points from the regression line (shown as solid horizontal lines in the diagram below) is made as small as possible.

Discussion point

How can you use your calculator, or a spreadsheet, to find the equation of the line of regression of x on y?

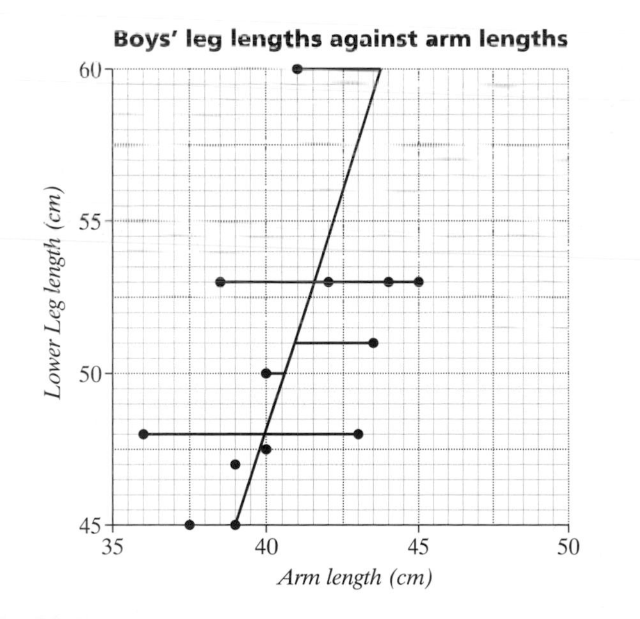

Note: This can be done easily using your graphic calculator or spreadsheet software by reversing the lists used for x, y.

It is probably best to redraw the scatter diagram with leg length on the horizontal axis and arm length on the vertical axis with a new regression line $y = 0.338x + 23.5$, where x is now leg length and y is now arm length.

Activity 2.3D

1 Use your graphic calculator or spreadsheet to confirm that the equation of the regression line to predict arm length, y, from leg length, x, is $y = 0.338x + 23.5$.

2 Draw a scatter diagram with leg length, x, on the horizontal axis. On this draw the new regression line.

3 **a** Use the regression line to estimate the arm length of a boy with a leg length of 51 cm.

b For what range of values of leg length is the regression line likely to give good estimates of arm length?

2.4 Revision summary

You can use scatter diagrams and correlation coefficients to give you an idea of the relationship between sets of **bivariate data**. If there is some **correlation** between the sets of data you can use a regression line to estimate the value of one variable if you know the value of the other. In this chapter, you are working only with linear models – you are seeing how closely data fits to a perfect straight line.

Scatter diagrams

You can use scatter diagrams to illustrate bivariate data; each point on the graph represents one item (a person, or a school, or a country, for example) for which you have two pieces of data such as a person's height and weight, a school's number of pupils and number of teachers, a country's area and population. The scatter diagram consists of points representing all the people or schools or countries.

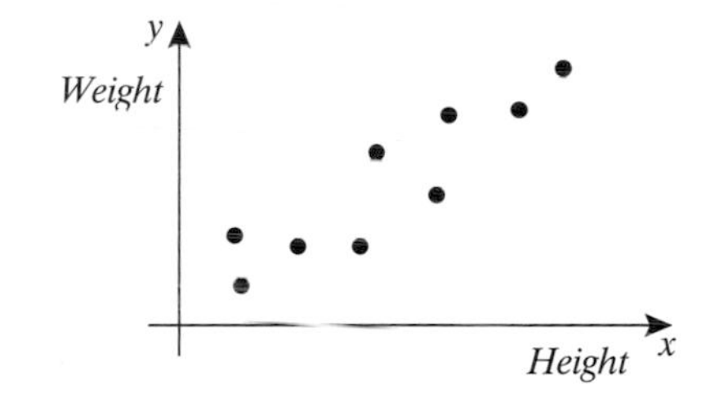

Each point has the value of one variable as its x co-ordinate and the corresponding value of the other variable as its y co-ordinate.

Points scattered all around the diagram indicate **no correlation** between the two sets of data; points grouped around a diagonal sloping upwards show **positive correlation** and points grouped around a diagonal sloping downwards indicate **negative correlation**.

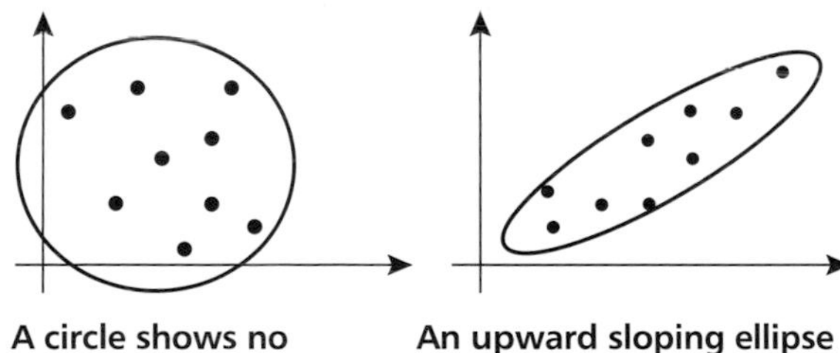

A circle shows no correlation.

An upward sloping ellipse shows positive correlation.

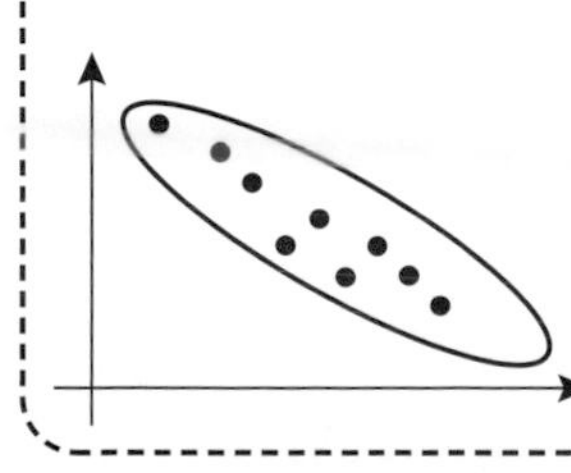

A downward sloping ellipse shows negative correlation.

Correlation coefficients

The strength of the relationship between the two variables can be measured by calculating a **correlation coefficient**. The one used in this course is **Pearson's product moment correlation coefficient**, r.

It is easiest to find this using your calculator by entering the data in two lists and then using the statistical functions.

There are a number of different formulae for finding r. These are quite complex. Three versions are

① $$r = \frac{N\Sigma XY - \Sigma X \Sigma Y}{\sqrt{(N\Sigma X^2 - (\Sigma X)^2)(N\Sigma Y^2 - (\Sigma Y)^2)}}$$

② $$r = \frac{S_{xy}}{\sqrt{S_{xx}S_{yy}}}$$

where $S_{xy} = \dfrac{\Sigma XY}{n} - \dfrac{\Sigma X \Sigma Y}{n^2}$,

$S_{xx} = \dfrac{\Sigma X^2}{n} - \left(\dfrac{\Sigma X}{n}\right)^2$, and

$S_{yy} = \dfrac{\Sigma Y^2}{n} - \left(\dfrac{\Sigma Y}{n}\right)^2$.

③ $$r = \frac{S_{xy}}{\sqrt{S_{xx}S_{yy}}}$$

where $S_{xy} = \Sigma(x - \bar{x})(y - \bar{y})$

$S_{xx} = \Sigma(x - \bar{x})^2$

$S_{yy} = \Sigma(y - \bar{y})^2$

If r is close to zero, there is **no correlation**.

If r is close to $+1$ there is **strong positive correlation** and if r is close to -1 there is **strong negative correlation**.

Regression lines

A line drawn on a scatter diagram showing the general trend of the points is called a line of best fit if drawn by eye and a regression line if calculated according to a particular method.

In either case it should pass through the mean point. The standard method used to find the regression line is the 'method of least squares', giving a regression line $y = ax + b$, where

$$a = \frac{N\Sigma XY - \Sigma X\Sigma Y}{N\Sigma X^2 - (\Sigma X)^2} \text{ and}$$

$$b = \frac{\Sigma X^2\Sigma Y - \Sigma X\Sigma XY}{N\Sigma X^2 - (\Sigma X)^2}$$

Again, it is easiest to find the regression line using the statistical functions of your graphic calculator.

$y = ax + b$ is the regression line of y on x, which can be used to estimate likely values of y given values of x.

2.5 Preparing for assessment

Your coursework portfolio

In this chapter you have learned how to investigate how one variable might be related to another using scatter diagrams and ideas of correlation and regression. You should now be able to use spreadsheet software and your graphic calculator to draw such diagrams, calculate correlation coefficients and find lines of regression (see *Using Technology*).

You may wish to incorporate such ideas into your coursework portfolio evidence. It is quite often easy to do so when you have data you have collected yourself. If working with secondary data you have to be careful that it is appropriate to work with these statistical techniques. It is, however, possible to do so as the work you did in **Activities 2.2C**, **2.3B** and others demonstrates.

Make sure that you interpret what you find in terms of **the situation you are investigating**. Points you should consider include:

- Does the scatter diagram indicate positive, negative or no correlation? What does this mean in terms of the two variables?
- What does the correlation coefficient indicate about the strength of any correlation?
- Would the situation change significantly if you removed any outliers from the data?
- Can you use the regression line to predict values for which you have no data?
- Is there a causal relationship between the two variables?

If you have used ideas of correlation and regression in your coursework portfolio, make sure that you have a graph showing the data with the regression line plotted carefully. You can do this by hand, on graph paper or using spreadsheet software.

Note: The Nuffield FSMQ website provides many datasets and other resources that may be of use to you.

Practice exam questions

Data

Resource Sheet 2.5

1 The table shows Key Stage 2 test information for a sample of 20 primary schools.

Percentages of pupils in sample of 20 primary schools gaining level 4 or above in Key Stage 2 tests

School number	Maths	Science	English
1	35	74	35
2	68	84	68
3	43	63	60
4	56	77	69
5	67	76	57
6	89	95	63
7	61	83	53
8	69	80	77
9	46	73	69
10	63	77	72
11	65	79	50
12	72	85	75
13	66	94	70
14	85	97	88
15	59	87	70
16	73	87	75
17	68	92	78
18	69	100	74
19	59	87	62
20	82	92	49

The data are shown in the two scatter diagrams. In the first, each point represents one of the schools and shows the percentage of pupils at that school gaining level 4 or above in English and in mathematics. A line of best fit is drawn on this diagram. In the second diagram, the percentages of pupils gaining level 4 or above in science and mathematics are shown at each of the 20 primary schools.

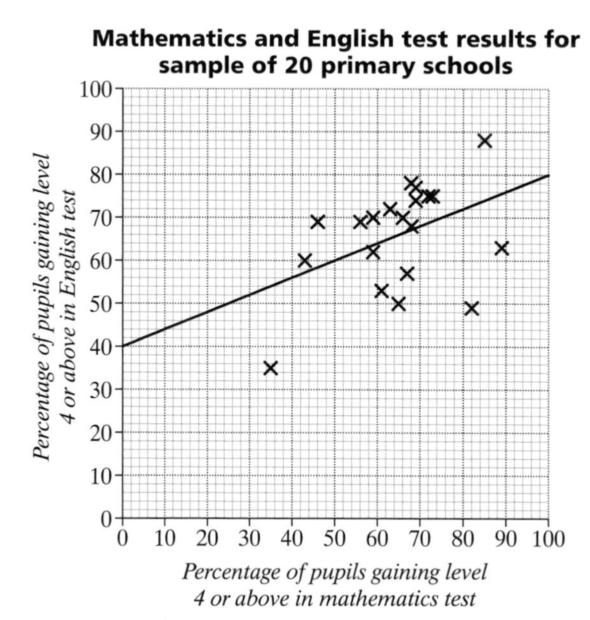

Mathematics and Science tests result for sample of 20 primary schools

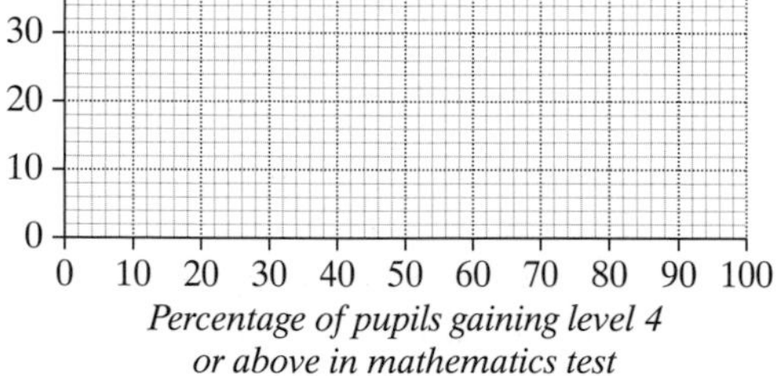

Question

a Use your calculator to find

i the mean value, $\bar{x}$, of the maths data

ii the mean value, $\bar{y}$, of the science data

iii the correlation coefficient

iv the coefficients a and b for the line of best fit, $y = ax + b$.

b Plot a line of best fit on a copy of the scatter diagram of the maths and science data.

c Use the graphs to estimate

i the percentage of pupils gaining level 4 or above in English in a school where 50% of pupils gain level 4 or above in mathematics

ii the percentage of pupils gaining level 4 or above in science in a school where 50% of pupils gain level 4 or above in mathematics.

d The correlation coefficient for the English and mathematics results is 0.430. Which of your two estimates in **c** do you consider to be the more accurate? Explain your answer.

e Compare and contrast the data shown in the two scatter diagrams.

Data

2 In an experiment, a person was asked to estimate the weight of some objects by comparing them with a 100 g weight.

Actual weight (grams)	Estimated weight (grams)
x	y
19	10
33	25
46	35
63	60
74	90
98	100
136	150
212	280

The data is also shown in the scatter diagram below.

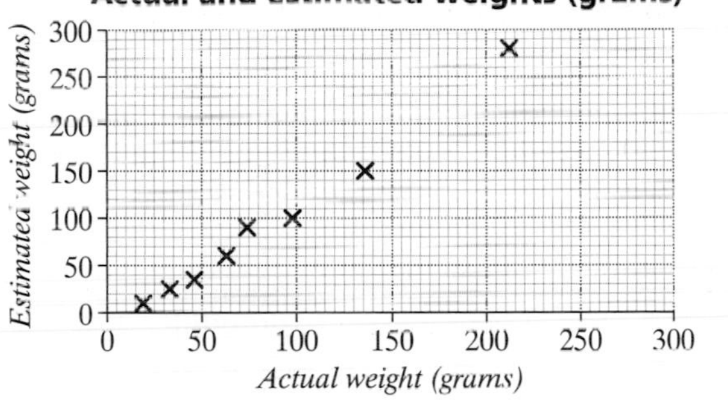

Question

a Use your calculator to find

i the mean value, $\bar{x}$, grams, of the actual weight

ii the mean value, $\bar{y}$, grams, of the estimated weight

iii the correlation coefficient

iv the equation of the line of best fit, $y = ax + b$.

b Draw the regression line on the scatter diagram. What does the line indicate that this person would estimate the weight of a 200 gram object to be?

c What information does the gradient of the line give you?

3 The Normal Distribution

Contents

What grade will you get when you take the examination for *Using and Applying Mathematics*? Well, of course, this will depend on a number of factors. As well as your mathematical ability it will also depend on things like how much time you spend working through this book, how much effort you put into developing your coursework portfolio and so on.

We can expect that when a large number of people take an examination, if we display the results as a **bar chart** or histogram it will look like the one on the right – it will have a 'bell shape'. This is because few students will get very high marks, few will get very low marks and most students will have a mark somewhere in the middle of the distribution. We expect the results to be *normally distributed*.

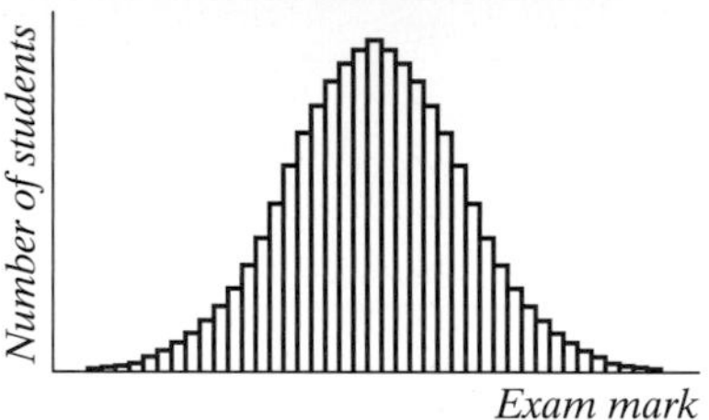

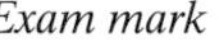

Some years ago exam grades were awarded in fixed proportions. So, for example, the bottom 30% of students always failed. Of course this was unfair – what if only the very best students were to take the exam? Using this method 30% of them would still fail.

For some time now exams of the type you will sit have been graded differently. There has been a move to make the exams more like the driving test. This has criteria you have to meet to pass. For example, in a driving test you will be asked to complete manoeuvres such as:

- reversing into a limited opening either to the right or left
- turning round by means of forward and reverse gears
- reverse parking in reverse gear
- reversing into a parking bay at the driving test car park.

It is not easy to have criteria like these for maths exams. But attempts are made by examiners to judge from year to year that those students who are awarded a pass grade are able to do the same sort of things. This means that, depending on the level of difficulty of the exam, the pass mark is set and the number of people passing depends on the performance of the students sitting the paper.

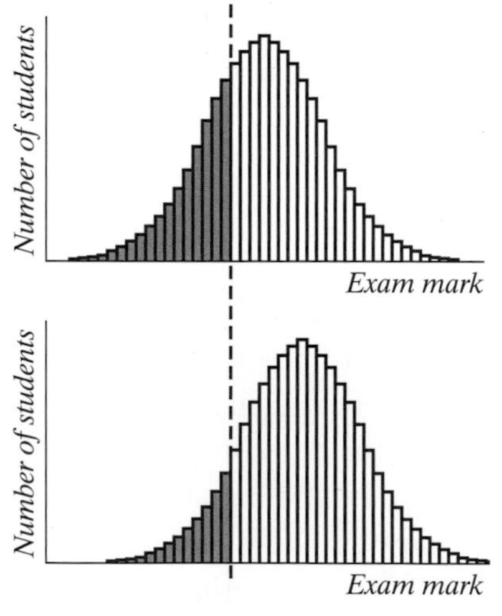

The number passing the exam depends on the ability of those sitting it.

The **normal distribution** is the most important and widely-used theoretical model in statistics. It was originally derived by the French mathematician De Moivre (1667–1754), but more extensive work was carried out later by the German mathematician Gauss (1777–1855). Physicists and engineers often refer to the normal distribution as 'the Gaussian distribution'.

You will learn how to use this important model in this chapter.

3.1 Samples from normal and other distributions

Computer workstation

Throughout this sub-section you will consider how a manufacturer designs a computer workstation for workers.

Poor ergonomic design can lead to eye, neck and back strain and wrist, hand, elbow and shoulder diseases among computer operators. One of the primary causes of these problems is the improper positioning of monitor screens. Ideally the top of the monitor screen should be at, or slightly below, eye level and as far away from the eyes as possible, bearing in mind that the operator needs to be able to read characters on the screen without eye strain. A good rule of thumb is that the monitor screen should be an arm's length away from the operator.

For a particular computer workstation, in which the monitor is to be a flat screen mounted on the wall above the keyboard, the height of the screen is to be adjustable so that it can be changed to suit most operators. Enough space is to be allowed for the majority of operators to sit an arm's length away from the screen.

In order to work out the dimensions of the computer workstation, measurements are taken from a sample of 1000 adults including 500 males and 500 females. Each adult is seated comfortably on a chair, then the height of their eyes above the floor is recorded. The length of their arms from shoulder to fingertips is also measured and recorded.

The results are summarised in the tables below.

Sitting eye height (mm)	No. of females	No. of males
1000–	4	0
1020–	12	0
1040–	15	0
1060–	28	0
1080–	50	3
1100–	69	4
1120–	89	12
1140–	80	23
1160–	64	36
1180–	36	47
1200–	27	56
1220–	16	72
1240–	7	88
1260–	3	57
1280–	0	47
1300–	0	25
1320–	0	18
1340–	0	9
1360–	0	3
1380–	0	0
Totals	500	500

Arm length (mm)	No. of females	No. of males
620–	3	0
640–	11	0
660–	41	0
680–	92	0
700–	132	2
720–	120	9
740–	69	27
760–	25	71
780–	6	114
800–	1	122
820–	0	89
840–	0	46
860–	0	15
880–	0	4
900–	0	1
Totals	500	500

Computer workstation

Features of normal distributions

Other distributions

Ergonomics is the science of fitting the job to the worker. It is concerned with the measurement of human factors, such as body dimensions and movements.

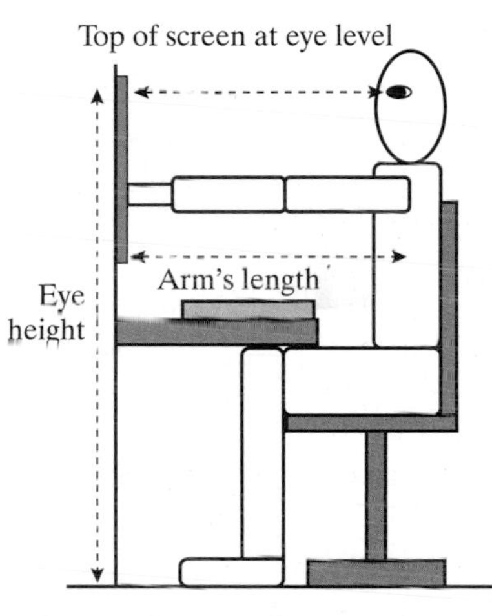

Discussion point

If this data is to be useful it should be taken from a *representative* sample of adults.
What does this mean?
How could it be achieved?

Note: When the classes are of equal width, the frequency can be used on the vertical axis of a histogram since the vertical axis has to be proportional to frequency density. Can you explain why?

Resource Sheet 3.1A

Activity 3.1A

Share this work and pool your results.

1 **a** Use the eye height data to draw histograms showing
  i male eye heights **ii** female eye heights
  iii adult eye heights.
 b Compare the histograms and comment on similarities and differences.

2 **a** Use the arm length data to draw histograms showing
  i male arm lengths **ii** female arm lengths
  iii adult arm lengths.
 b Compare the histograms and comment on similarities and differences.

3 Calculate estimates of the mean and standard deviation of
 a male eye heights **b** female eye heights
 c adult eye heights **d** male arm lengths
 e female arm lengths **f** adult arm lengths.

4 Write down the modal class and estimate the median of
 a male eye heights **b** female eye heights
 c adult eye heights **d** male arm lengths
 e female arm lengths **f** adult arm lengths.

5 Use the results from **questions 3** and **4** to compare the eye heights and arm lengths of men with those of women.

Use frequency density on the vertical axis.

Add the frequencies in the male and female columns to give the total number of adults in each group.

Keep your histograms and the results of your calculations – you will need them later in this section.

Reminder: The median can be estimated using linear interpolation or a cumulative frequency graph.

The manufacturer wants to make the height of the monitor screen adjustable for the people between the height needed by a small female to that needed by a large male. To estimate this height range the eye height of the 25th person in the female sample and the 475th person in the male sample are to be used.

It is common practice to cater for all but the smallest 5% of females and the largest 5% of males.

The histogram below shows the female eye heights when seated. The area of each bar represents the frequency of the group. The total number of females in the first two groups is 16 (= 4 + 12).

These frequencies can be found in the frequency table or worked out from the histogram:
Frequency of 1st group
$= 20 \times 0.2 = 4$
Frequency of 2nd group
$= 20 \times 0.6 = 12$

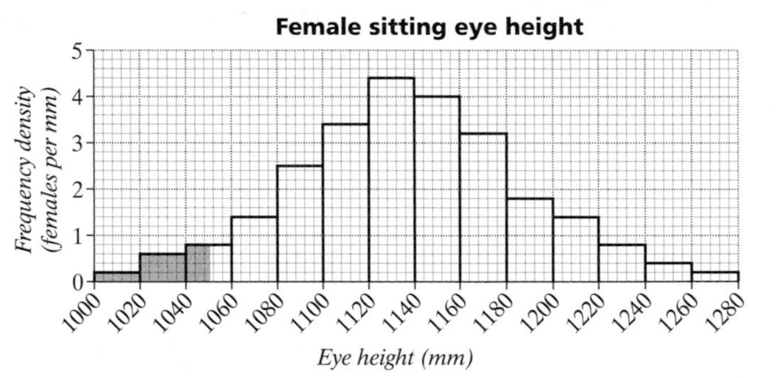

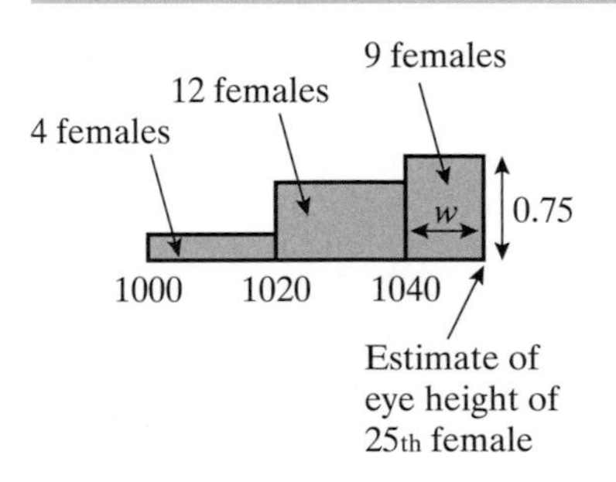

The shaded area in the histogram represents 25 females. The 25th female is the 9th person in the third group and so the shaded area in the 3rd group represents 9 females.

Since the height of 3rd shaded rectangle is 0.75 (the frequency density of this group), its width is $w = 9 \div 0.75 = 12$.

Estimate of the eye height of the 25th female $= 1040 + 12$
$= 1052$ millimetres.

Discussion points

Can you explain this calculation? What assumptions are being made by estimating the eye height of the 25th female in this way?

Activity 3.1B

Use your histograms from **Activity 3.1A**.

1 a Estimate the eye height of the 475th male.
  b The height of the top of the monitor screen is to be adjustable to suit all from the 25th female to the 475th male. Write down
    i the minimum height
    ii the maximum height
    iii the range of heights
  that will be catered for.

2 a Estimate the arm length of
    i the 25th female
    ii the 475th male.
  b Estimate the arm length of the adult who is
    i 5% of the way through the sample of all adults
    ii 95% of the way through the sample of all adults.
  c State the range of the arm length distance the manufacturer will use when designing the computer workstation.

Discussion points

How could the eye height of the 25th female and the 475th male be estimated from the original frequency tables instead of the histograms?

Why is the person 5% of the way through the adult population not the same as the person 5% of the way through the female population?

Nuffield Resource Starter – 'Stature'

Features of Normal Distributions

Look again at the histograms of eye heights you have drawn in **Activity 3.1A**. The histograms for the female and male eye heights are both reasonably symmetrical, reaching a peak near the centre of the distribution. This shape of histogram often occurs when the underlying distribution is normal.

Discussion point

Can you think of any other variables that are likely to give reasonably symmetrical histograms like these with peaks near the centre of the distribution?

A **normal distribution** is a theoretical model of the whole population. It is perfectly symmetrical about the central value. The sketch below shows in theory what a histogram of a representative sample from such a distribution is expected to look like.

The population consists of **all** the members of the group being studied.

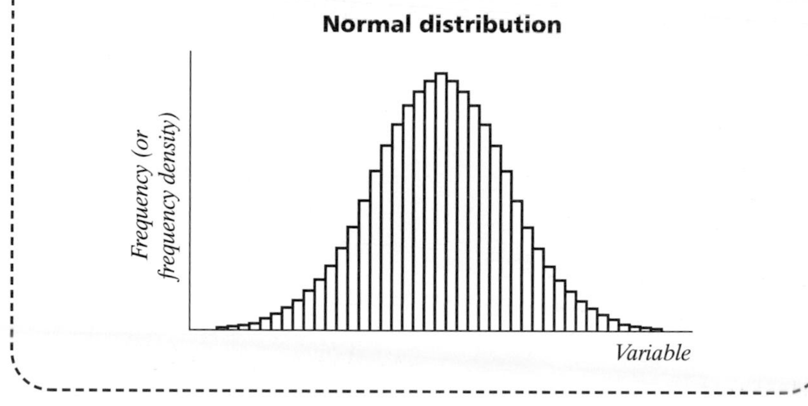

Discussion points

Why does symmetry imply that the mean, mode and median values are equal?

In practice, samples from a normal distribution can be quite variable. Compare the other histograms you have drawn in **Activity 3.1A** with this shape.

Were the mean, mode and median values approximately equal?

Perfect symmetry means that in theory the mean, mode and median of a normal distribution are equal and lie at the centre of the distribution.

In practice it is very rare to get such perfect symmetry from a random sample, but many distributions give histograms that are similar to that shown in the sketch. Variables that give such distributions include height, weight, time taken to perform a task and so on. The normal distribution provides a good model of such distributions.

There are other features that samples from normal distributions usually have.

Activity 3.1C

1. Take measurements from a sample of males or females of a similar age. The measurements could be lengths such as eye height when sitting and arm length as in the previous activity or other continuous variables such as weight or time taken to perform a task. The larger the sample you use the better. Work in a group if possible with each member of the group collecting data which you share.
2. Use your results to draw histograms and compare them with the theoretical shape of a sample from a normal distribution (shown above).

Discussion point

Why is it better to use a large sample rather than a small sample?

Computer simulations of samples from normal distributions are available on the Internet. Use one of these to look at the variation in histogram shapes that can arise in such samples. Also compare mean values if the simulation allows you to do this.

For the sample of female eye heights the mean was 1137 mm and the standard deviation was 49 mm (nearest mm).

The mean minus one standard deviation is $1137 - 49 = 1088$ mm and the mean plus one standard deviation is $1137 + 49 = 1186$ mm.

The shaded area in the histogram below represents the number of females with eye heights between 1088 and 1186 i.e. within ± 1 standard deviation of the mean.

Sitting eye height (mm)	No. of females	Frequency density
1000–	4	0.2
1020–	12	0.6
1040–	15	0.75
1060–	28	1.4
1080–	50	2.5
1100–	69	3.45
1120–	89	4.45
1140–	80	4
1160–	64	3.2
1180–	36	1.8
1200–	27	1.35
1220–	16	0.8
1240–	7	0.35
1260–	3	0.15
1280–	0	0

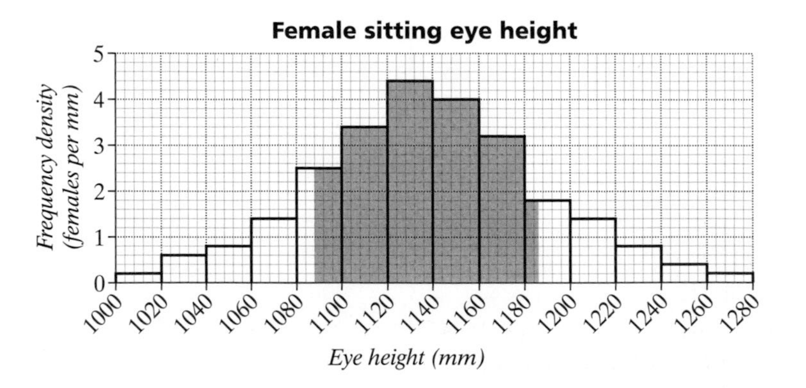

The area of the first shaded rectangle is $12 \times 2.5 = 30$.

This represents the 30 females with eye heights between 1088 mm and 1100 mm.

The area of the last shaded rectangle is $6 \times 1.8 = 10.8$.

This gives an estimate of approximately 11 females with eye heights between 1180 mm and 1186 mm.

Hence an estimate for the number of females within one standard deviation of the mean is $30 + 69 + 89 + 80 + 64 + 11 = 343$.

The proportion of females within one standard deviation of the mean is about 69% ($= \frac{343}{500} \times 100$). This is typical of a sample from a normal distribution.

Discussion point
Can you explain these calculations?

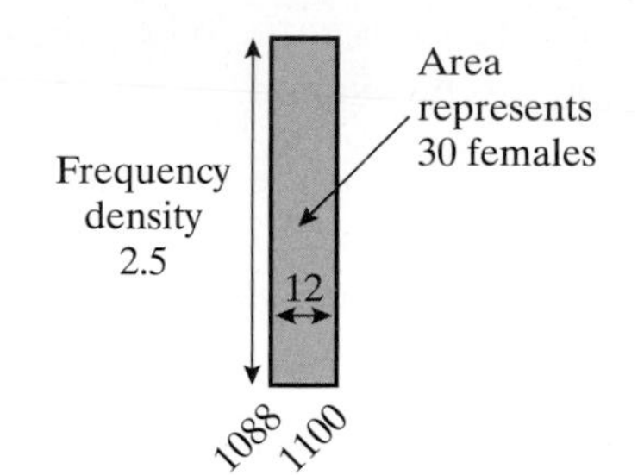

Usually in samples from normal distributions

- about two-thirds of the sample lie within one standard deviation of the mean
- about 95% of the sample lie within two standard deviations of the mean
- about 99% of the sample lie within three standard deviations of the mean.

Activity 3.1D

Use histograms from **Activities 3.1A** and/or **3.1C**.

In each case find the proportion of the sample that lies within one, two and three standard deviations of the mean. Compare your results with the proportions given above.

Other distributions

There are many other types of distribution. One of these is the **uniform distribution**. In theory this occurs when all values of the variable are equally likely to occur.

What shape of histogram would you expect to get from a sample from this distribution? Draw a sketch.

Like the normal distribution, the uniform distribution (sometimes called the **rectangular distribution**) is symmetrical. Many other distributions are not and it is useful to have a way of describing this lack of symmetry.

Discussion point
Would you expect the mean, mode and median values to be equal in a sample from a uniform distribution?

Discussion point
Are the mean, mode and median values equal in *any* perfectly symmetrical distribution?

☞ Subtracting the median from the mean provides a rough and ready measure of symmetry. For a symmetrical distribution the difference is zero. When the distribution is not symmetrical it is said to be **skewed**.

If the mean is greater than the median, subtracting the median from the mean gives a positive value and the distribution is said to be **positively skewed**. The peak value of such a distribution is near the start of the distribution.

A **negatively skewed** distribution occurs when subtracting the median from the mean gives a negative value. In this case the peak value is near the end of the distribution.

Examples of skewed distributions are shown in the sketches below.

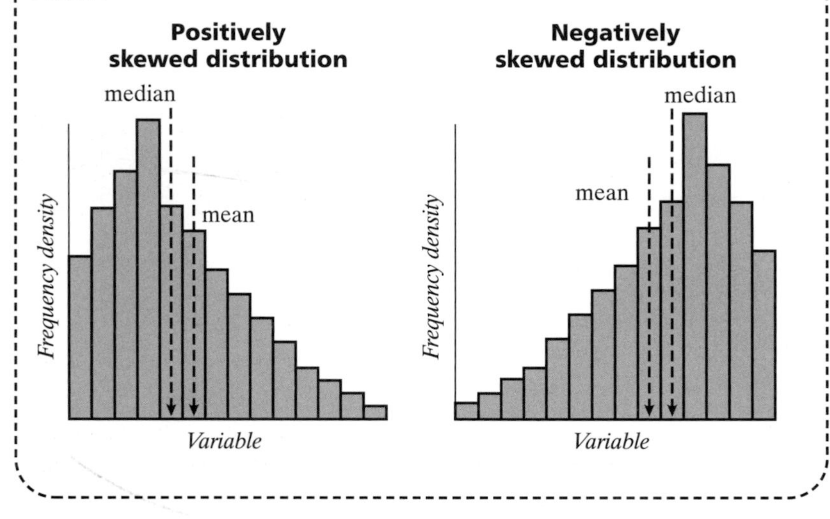

Note: Because the median is the value halfway through the distribution, it divides the area under the histogram into two equal halves.

Sometimes a distribution may have more than one peak value. When the male and female seated eye heights are combined into a single sample of adult eye heights, the histogram is as shown below.

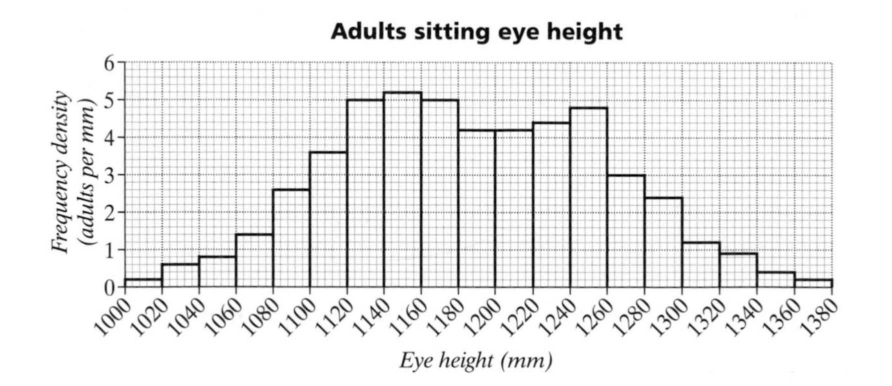

Discussion point

What other histogram that you have drawn has shown a bi-modal distribution?

This distribution has two peaks. It is **bi-modal**.

Some distributions have more than two modes and are said to be **multi-modal**.

Activity 3.1E

1 The tables below show data collected at a weather station in the South East of England over the course of a year.

Sunshine (hours)	No. of days
0–	90
1–	35
2–	34
3–	29
4–	34
5–	20
6–	26
7–	14
8	23
9–	17
10–	14
11–	12
12–	3
13–	9
14–	5
15–	0

Rainfall (mm)	No. of days
0–	131
0.2–	82
0.4–	24
0.6–	15
1.0–	24
2.0–	29
3.0–	8
4.0–	19
6.0–	12
8.0–	5
10.0–	4
12.0–	3
14.0–	3
16.0–	2
18.0–	4

a **i** Draw a histogram to illustrate the sunshine data.
ii Describe the shape of the histogram and explain what it shows.

b **i** Draw a histogram to illustrate the rainfall data.
ii Describe the shape of the histogram and explain what it shows.

2 The table shows the prices of the Ford Fiestas for sale in the *Gloucestershire Car Finder Edition Number 1040*.

Price (£)	Frequency
0–	8
1000–	19
2000–	45
3000–	25
4000–	23
5000–	9
6000–	11
7000–	1

a **i** Draw a histogram to illustrate the data.
ii Describe the shape of the histogram.

b Find
i the mean and standard deviation of the prices
ii the median price
iii the difference mean − median.
What does this suggest about the symmetry of the distribution?

c Use your histogram to estimate the number of Ford Fiestas advertised in this edition of *Car Finder* for a price
i less than £4200;
ii within one standard deviation of the mean;
iii Express your estimate of the number of Ford Fiestas within one standard deviation of the mean as a percentage of the total. Compare this with expected proportion for a normal distribution of two-thirds.

Remember to use **frequency density** on the vertical axis of the histogram.

Discussion point

What problems do you encounter when you attempt to draw a histogram of the rainfall data? How can you overcome these problems?

If you have time, investigate the distributions of the prices of other types of cars. Try different ways of grouping the prices and compare the resulting histograms.

You could also investigate house prices. What would you expect a histogram of house prices to look like?

3 The table shows the times taken by the runners in the London Marathon in 2002.

Time (h:min)	No. of men	No. of women
2:05–2:29	54	11
2:30–2:44	202	4
2:45–2:59	867	30
3:00–3:14	1396	100
3:15–3:29	1958	241
3:30–3:44	2616	436
3:45–3:59	3491	672
4:00–4:14	2730	713
4:15–4:29	3052	992
4:30–4:44	2584	1087
4:45–4:59	2071	963
5:00–5:14	1283	700
5:15–5:29	921	550
5:30–5:44	625	476
5:45–5:59	391	265
6:00–6:14	226	188
6:15–6:29	153	164
6:30–6:44	95	115
6:45–7:14	53	61

a Using the same scale, draw histograms to show the time taken by

i the men **ii** the women **iii** all the runners.

b Subtract the *median* from the *mean*

i the men **ii** the women **iii** all the runners.

c Describe the shape of each histogram and briefly compare them.

d Find the mean time for

i the men **ii** the women **iii** all the runners.

e Find the median time for

i the men **ii** the women **iii** all the runners.

f Use your histogram of men's times to estimate the number of men who were faster than the mean men's time.

g **i** Use your histogram of women's times to estimate the number of women who were slower than the mean women's time.

ii Why would use of the histogram not give an accurate value for the number of women who beat the 2001 women's best time of 2 hours 24 minutes?

Discussion point

The times were rounded to the nearest minute. What are the class boundaries?

Discussion points

Try to explain the underlying reasons why the distributions give the shapes of the histograms you have drawn for each question in this activity.

4 The times between successive cars arriving at a petrol station were measured and recorded. The results are given in the table below.

Time (seconds)	No. of cars
$0 \leqslant t < 30$	393
$30 \leqslant t < 60$	239
$60 \leqslant t < 90$	145
$90 \leqslant t < 120$	88
$120 \leqslant t < 150$	53
$150 \leqslant t < 180$	32
$180 \leqslant t < 210$	20
$210 \leqslant t < 240$	12
$t \geqslant 240$	18

a What is the modal class?

b Estimate the median time.

c Taking the upper bound of the last group to be 5 minutes:

i draw a histogram to illustrate the data and describe its shape

ii calculate an estimate of the mean time

iii state whether you think the estimated mean or estimated median gives the better measure of location and explain why.

This type of distribution is called an **exponential distribution**. Can you explain why?

5 **a** Use your calculator to generate a sequence of 100 random digits.

b **i** Draw a bar chart to show the distribution of digits in your sample.

ii Is the shape of the bar chart what you would expect? Explain your answer.

c Share the work for this part with other students. Repeat **parts a** and **b** with larger samples of random numbers. Compare and comment on the results.

On graphic calculators the random number generator is usually listed in the probability menu as RAN or RAND.

Note that this is an example of a **discrete uniform distribution**.

3.2 The standard normal distribution

Probability density functions

The standard normal distribution

The work you have been carrying out in this book has involved collecting, illustrating, summarising and interpreting real data. This branch of the subject is usually called **descriptive** statistics.

In this section you will start to study **inferential** statistics. This involves making predictions about a population based on a theoretical model of it. You will use the theoretical normal model for data such as that you have just met in **Section 3.1**.

Probability density functions

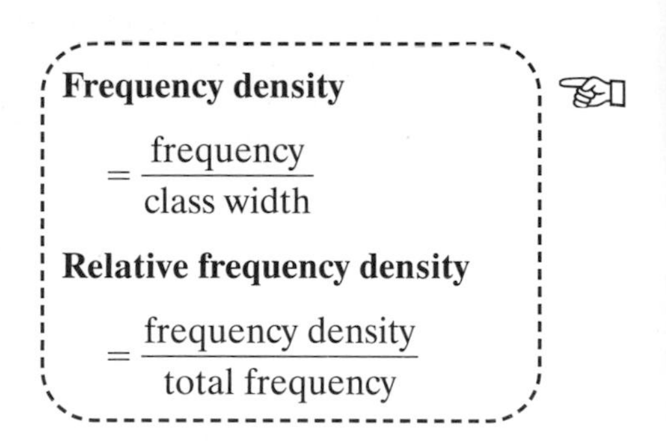

Frequency density
$= \dfrac{\text{frequency}}{\text{class width}}$
Relative frequency density
$= \dfrac{\text{frequency density}}{\text{total frequency}}$

In **Section 3.1** you investigated samples from normal and other distributions using histograms. The area under a histogram is proportional to frequency. When data groups have different widths it is best to use **frequency density** on the vertical axis. The area of each bar of the histogram then gives you frequency. Frequency density can also be used when the groups have equal widths.
If the frequency density of each data group is divided by the total frequency, the result is called the **relative frequency density**.
A histogram drawn using relative frequency densities has the same shape as a histogram using frequency density, but now the area under the histogram gives the *proportion* of the sample rather than the number of readings. The total area then gives a value of 1.

Activity 3.2A

Resource Sheet 3.2A

1 The female eye height data from the beginning of **Section 3.1** has been used to start a relative frequency density table below.

Sitting eye height (mm)	Female frequency	Frequency density	Relative frequency density
1000–	4	0.2	0.0004
1020–	12	0.6	0.0012
1040–	15	0.75	0.0015
1060–	28	1.4	0.0028
1080–	50		
1100–	69		
1120–	89		
1140–	80		
1160–	64		
1180–	36		
1200–	27		
1220–	16		
1240–	7		
1260–	3		
Total	500		

a Copy and complete this table.

b Draw a histogram using **relative frequency density** on the vertical axis.

Since each class width is 20, frequency density

$$= \frac{\text{frequency}}{20}$$

The frequency density is the number of females per mm.

The relative frequency density for each group

$$= \frac{\text{frequency density}}{500}$$

You may notice a short cut: relative frequency density

$$= \frac{\text{frequency}}{10\,000}$$

This sort of short-cut only occurs when the class widths are all equal.

2 a Draw a relative frequency density histogram using the male eye height data from the beginning of **Section 3.1**.

  b Compare your histogram with that for the female eye heights. Describe the main similarities and differences.

3 Compare your histograms of female and male eye heights with the histograms you drew in **Activity 3.1A**.

If you have time, draw relative frequency density histograms of other sets of data from **Section 3.1**.

Each histogram drawn using relative frequency density has the same shape as the corresponding histogram drawn using frequency density, but now the area under the histogram gives the *proportion* of the sample rather than the number of readings.

The total area under a relative frequency density histogram is 1.

The eye height histograms are approximately symmetrical with the peak near the centre. This suggests that the samples are from normal distributions. If it was possible to measure the eye heights of the whole adult female population we could draw a relative frequency density histogram with many more groups giving a histogram like that shown in the graph.

Discussion point
You can obtain a relative frequency density histogram by rescaling a histogram that has frequency density on the vertical axis. Alternatively, you can consider it as a frequency density histogram that has been stretched by a scale factor $\frac{1}{k}$ where k is the total frequency. Make sure you can explain these statements.

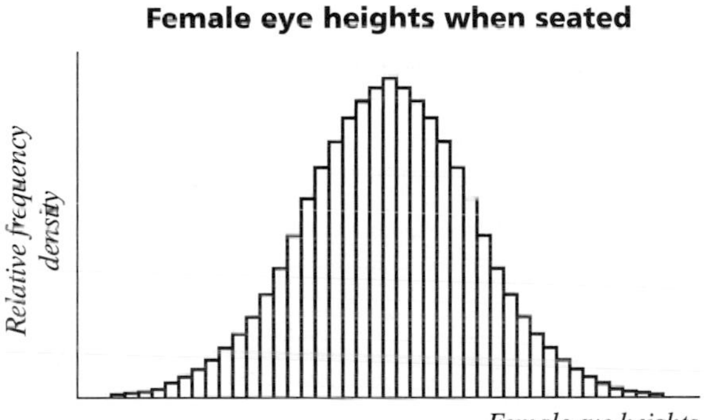

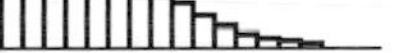

Discussion point
Can you explain why the total area under a relative frequency density histogram is 1?

The more groups there are, the closer the tops of the columns would approximate to a bell-shaped curve. The total area under the histogram would still be 1.

A **probability density function** (often abbreviated to pdf) is a theoretical model of such a curve. The female eye height is called a **random variable** (or **variate**). Random variables are often denoted by upper case letters, e.g. X.

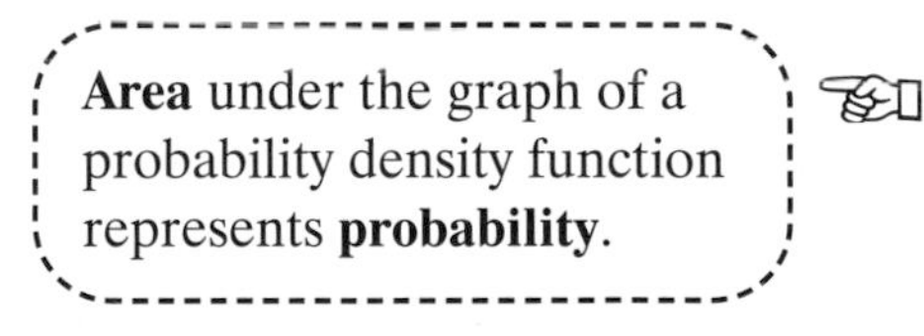
Area under the graph of a probability density function represents **probability**.

The graph of the probability density function for female eye heights is shown below. Note the graph is symmetrical about the mean value of 1140 cm and virtually the whole distribution lies within about three standard deviations (roughly 150 cm) of the mean, i.e. between 990 cm and 1290 cm.

Discussion point
What does this mean in terms of the female population?

The probability density function in this case is known as the normal distribution. It can be used to model data in many situations – particularly when it is naturally occuring and continuous as in this case.

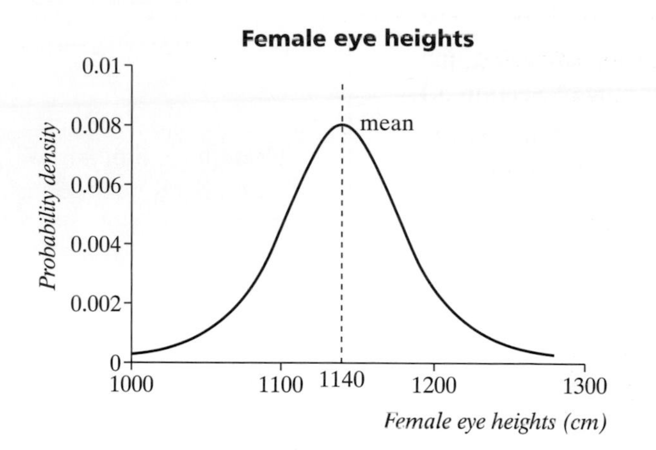

Discussion point

Compare this model with your relative frequency density histogram for female eye heights.

Before using the normal distribution to model real data you will investigate some of its properties.

The mean affects where the normal curve is positioned along the x axis. The standard deviation affects the width and height of the curve. If the standard deviation had been 100 cm instead of 49.3 cm, the curve would be wider, but less tall. This is because the total area under the curve has to remain 1.

Activity 3.2B

1 The sketches below show normal distributions with different means and standard deviations, drawn using approximately the same scale on each axis. Check that in each case the curve is symmetrical about the mean and that virtually the whole distribution lies within three standard deviations of the mean.

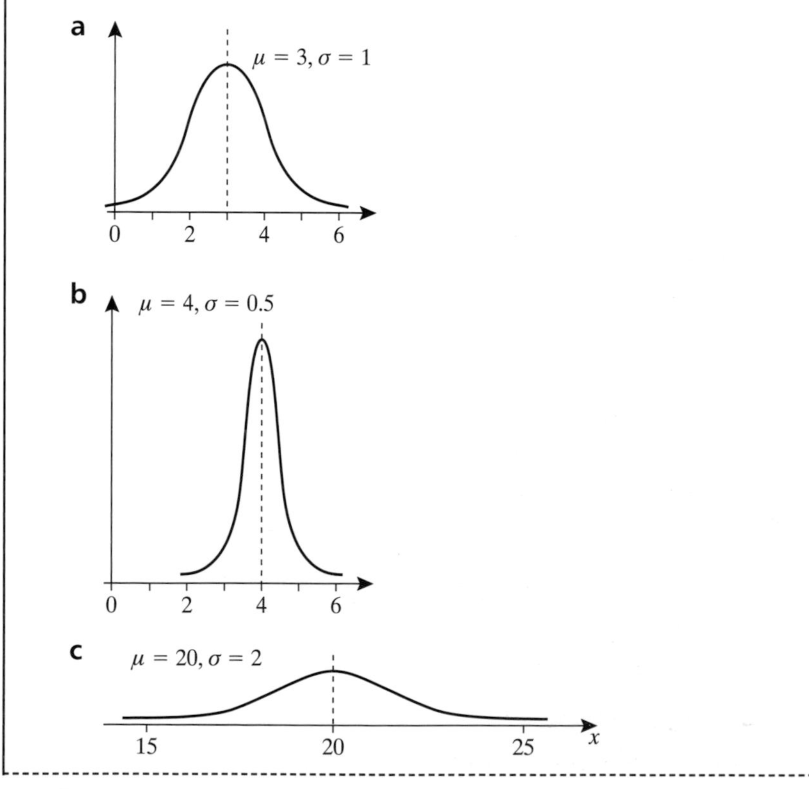

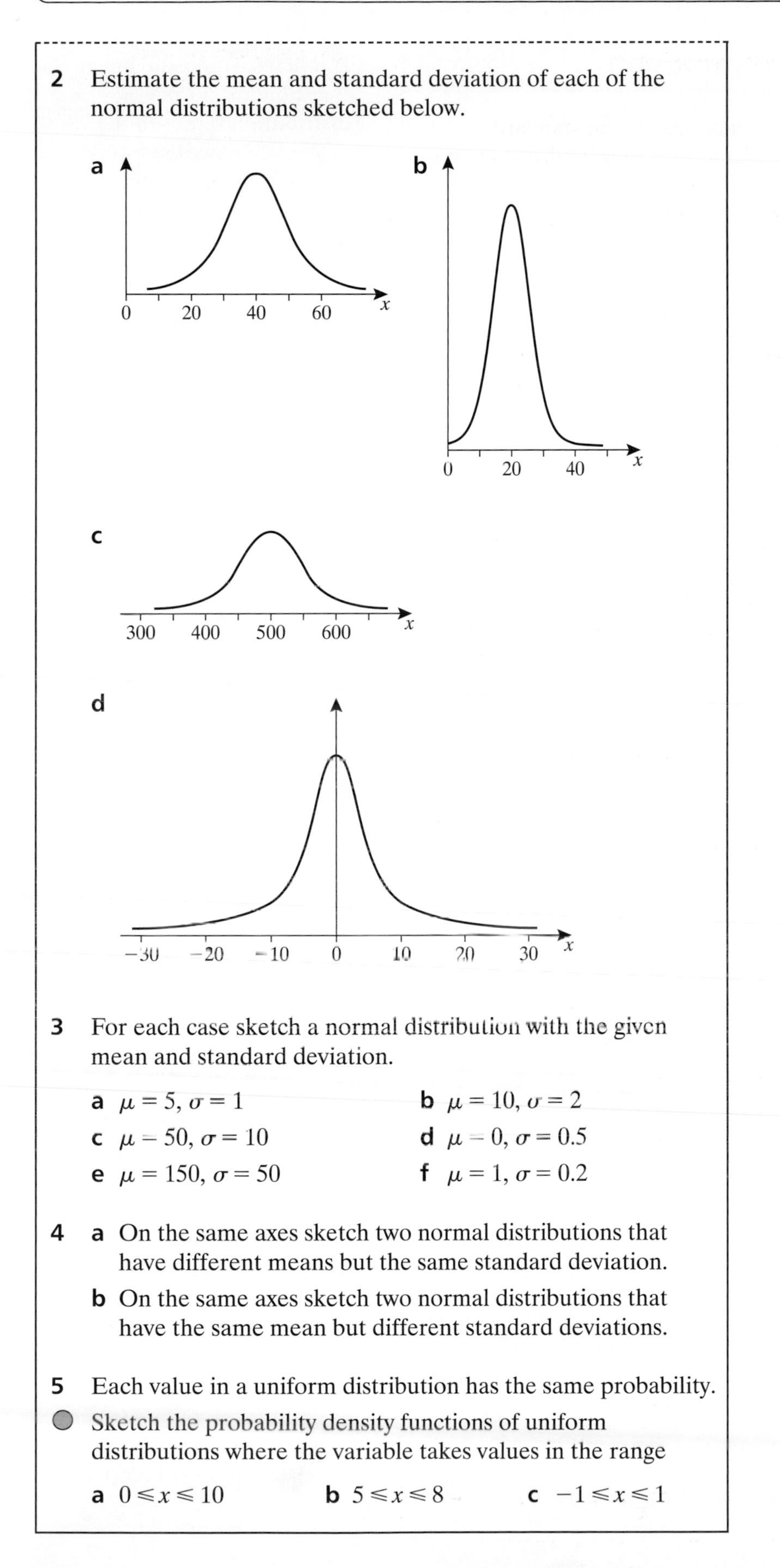

2 Estimate the mean and standard deviation of each of the normal distributions sketched below.

a **b** **c** **d**

3 For each case sketch a normal distribution with the given mean and standard deviation.

a $\mu = 5, \sigma = 1$

b $\mu = 10, \sigma = 2$

c $\mu = 50, \sigma = 10$

d $\mu = 0, \sigma = 0.5$

e $\mu = 150, \sigma = 50$

f $\mu = 1, \sigma = 0.2$

4 **a** On the same axes sketch two normal distributions that have different means but the same standard deviation.

b On the same axes sketch two normal distributions that have the same mean but different standard deviations.

5 Each value in a uniform distribution has the same probability. Sketch the probability density functions of uniform distributions where the variable takes values in the range

a $0 \leqslant x \leqslant 10$

b $5 \leqslant x \leqslant 8$

c $-1 \leqslant x \leqslant 1$

The standard normal distribution

The **standard** normal distribution has mean 0 and standard deviation 1. Initially this may not seem a very useful distribution to study. However, all normal distributions can be related to this distribution using ideas about transformations of functions that are explained in **Chapter 5** of **Algebra and graphs**, and you can therefore use it as the basis of all further theoretical analysis. The graph below shows the standard normal distribution. The letter z is usually used when working with the standard normal distribution.

If it helps, you could imagine the standard normal distribution representing errors when people estimate a length. Some people will estimate the length as being too long, giving a positive error, x, and some estimate it as too short, giving a negative error. The mean error is 0.

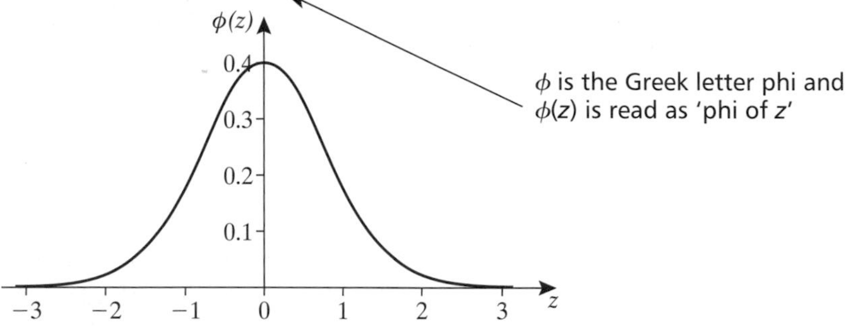

Note: $\phi(z) = \dfrac{1}{\sqrt{2\pi}} e^{-\frac{1}{2}z^2}$

You do not need to know this function but you may like to plot it on your graphic calculator.

The **area under the graph** between two values of Z gives the **probability** that a value taken at random from the standard normal distribution lies between them.

Because the pdf $\phi(z)$ s such a difficult function, it is not easy to calculate areas using it. Instead you can look up area values in a *Standard Normal Table*, or find them using your graphic calculator. You can find the *Standard Normal Table* that you should use on this course at the back of the book.

Φ is the capital letter phi. The function $\Phi(z)$, is called the **cumulative distribution function** of Z.

It gives $P(Z \leqslant z)$, i.e. the probability that a random value from the standard normal distribution is less than or equal to the particular value, z.

To look up a value of z in the table, use the first column for the whole number and first decimal place and then go across the table for the second decimal place.

For example, to look up $z = 1.17$, go down the first column to 1.1, then across the table to the column headed 0.07. You should find the value given in the table is 0.8790.

This can be written as $\Phi(1.17) = 0.8790$ and means that $P(Z \leqslant 1.17) = 0.8790$, i.e. the probability that a random value from the standard normal distribution is less than or equal to 1.17 is 0.879.

Note that $P(Z < 1.17)$ is also taken to be 0.8790.
Because the distribution is continuous, there are an infinite number of possible values of z. The probability of getting ***exactly*** 1.17 is so small that it is taken to be zero and it makes no difference whether 1.17 is included or not.

Since the total area under the curve is 1 (total probability = 1), the smaller unshaded area under the curve can be found by subtracting 0.879 from 1.

This gives $P(Z > 1.17) = 1 - 0.879 = 0.121$.

The table gives only positive values of z. Probabilities for negative values of z can be found using the symmetry of the curve.

For example, the sketches below show that
$P(Z \geqslant -0.82) = P(Z \geqslant 0.82) = 0.7939$ (from the table).

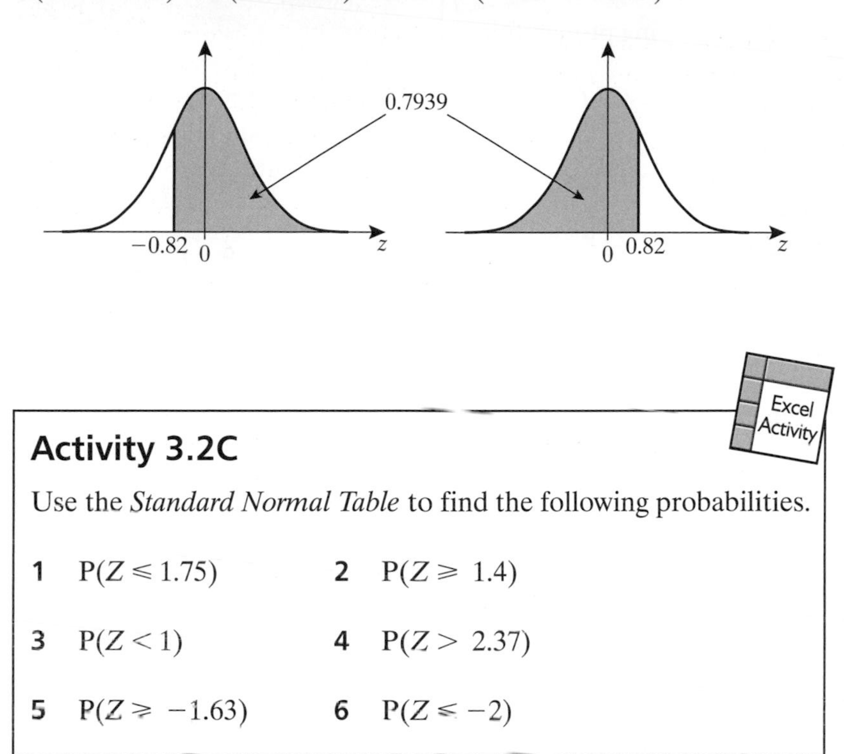

Activity 3.2C

Use the *Standard Normal Table* to find the following probabilities.

1 $P(Z \leqslant 1.75)$ 2 $P(Z \geqslant 1.4)$

3 $P(Z < 1)$ 4 $P(Z > 2.37)$

5 $P(Z \geqslant -1.63)$ 6 $P(Z \leqslant -2)$

Always draw a sketch showing the area you need to find.

After some practice you will probably not need to draw the reflection.
The first sketch is enough to show that the area is greater than 0.5. It can be found immediately from the table by looking up 0.82.
If the shaded area is the smaller part, take the value given in the table from 1.

Reminders: Draw a sketch for each area.

It does not make any difference to the method whether or not 'equals' is included in the inequality, i.e. the method is the same for $P(Z < 1)$ as for $P(Z \leqslant 1)$.

The mean value, 0, divides the area into two equal halves.

This fact can be used to give some probabilities.

For example, $P(0 \leqslant Z \leqslant 0.82) = 0.7939 - 0.5 = 0.2939$.

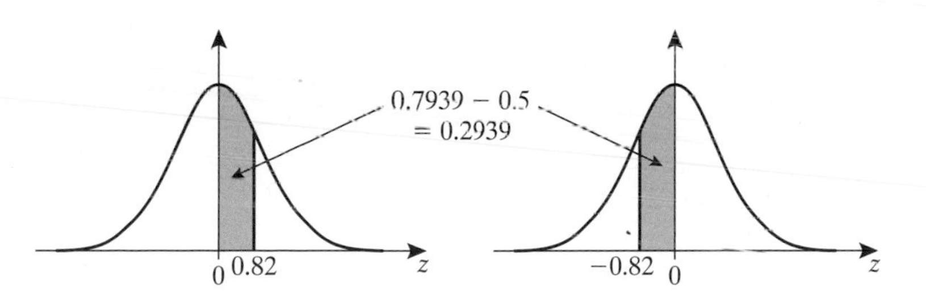

By symmetry $P(-0.82 \leqslant Z \leqslant 0)$ is also 0.2939.

Other probabilities can be found using more than one value from the table.

For example, $P(0.5 \leqslant Z \leqslant 2.5)$ is the area below 2.5 minus the area below 0.5.

$$\begin{aligned}P(0.5 \leqslant Z \leqslant 2.5) &= \Phi(2.5) - \Phi(0.5)\\ &= 0.9938 - 0.6915\\ &= 0.3023\end{aligned}$$

Activity 3.2D

Use the *Standard Normal Table* to find the following probabilities.

1 $P(0 \leqslant Z \leqslant 3)$

2 $P(-1.36 \leqslant Z \leqslant 0)$

3 $P(1.5 \leqslant Z \leqslant 2.4)$

4 $P(-2.05 \leqslant Z \leqslant -1.35)$

5 $P(-1 \leqslant Z \leqslant 1)$

6 $P(-2 \leqslant Z \leqslant 2)$

7 $P(-3 \leqslant Z \leqslant 3)$

8 $P(-0.6 \leqslant Z \leqslant 2.4)$

9 $P(-1.72 \leqslant Z \leqslant 0.4)$

10 $P(-0.26 \leqslant Z \leqslant 0.84)$

Reminder: Draw a sketch for each area.

Practice Sheet: Normal Distributions

Most graphic calculators are programmed to give probabilities for the standard normal distribution. For example, the diagram below shows how the answer for **question 3** above is given on one graphic calculator.

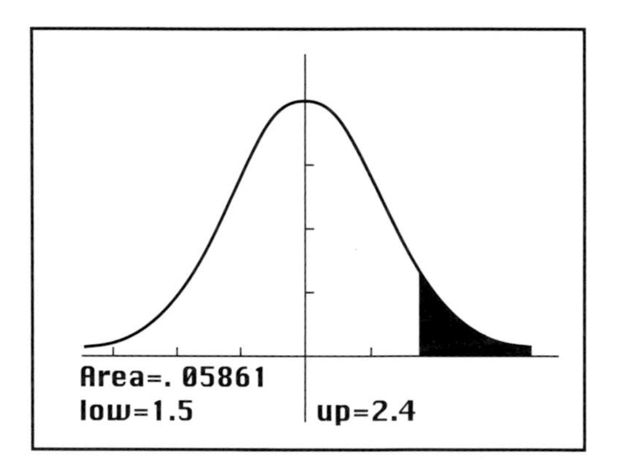

Activity 3.2E

Find out whether your graphic calculator gives normal probabilities. If so, learn how to use it by checking your answers for **Activities 3.2C** and **3.2D**.

3.3 Other normal distributions

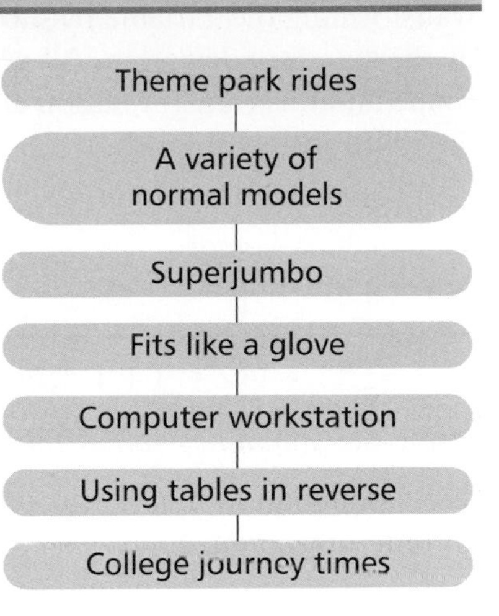

Theme park rides

Theme parks often have height restrictions on rides. On some rides these are *minimum* height restrictions imposed for safety reasons. On other rides the restrictions are *maximum* heights because the rides are intended for use by young children and the restriction is imposed to keep older children off them.

Before introducing height restrictions, theme park managers need to assess how they are likely to affect their customers. To do this they need information about the distribution of heights of children of different ages.

The table below gives estimates of the means and standard deviations of the distributions of heights of boys and girls with ages ranging from 5 years to 15 years.

	Boys		Girls	
Age	Mean (cm)	SD (cm)	Mean (cm)	SD (cm)
5	112.4	4.34	111.5	4.30
6	119.3	4.13	118.7	4.32
7	124.9	4.74	123.9	4.32
8	130.9	5.12	129.7	4.82
9	135.7	4.90	135.7	5.23
10	141.8	5.51	141.5	5.47
11	146.9	5.68	148.7	6.61
12	152.3	6.70	154.2	5.98
13	158.8	7.00	158.4	5.38
14	165.3	7.84	161.3	5.43
15	171.9	6.48	163.2	5.27

Suppose it is planned to impose a maximum height restriction of 1.5 metres on the Octopus and a minimum height restriction of 1.1 metres on the Calgary Stampede. The theme park manager wants to know how these are likely to affect children of different ages.

Consider, for example, 10-year-old boys. Assume that the height of boys of this age, X, is a normal variable with the mean and standard deviation given in the table above, i.e. $\mu = 141.8$ cm and $\sigma = 5.51$ cm. The probability that a boy of this age taken at random will be allowed to go on the Octopus is $P(X \leq 150)$.

Note: The height restriction of 1.5 metres must be converted to centimetres to match the units used for μ and σ.

This is given by the area under the normal curve up to 150 cm, as shown in the sketch opposite.

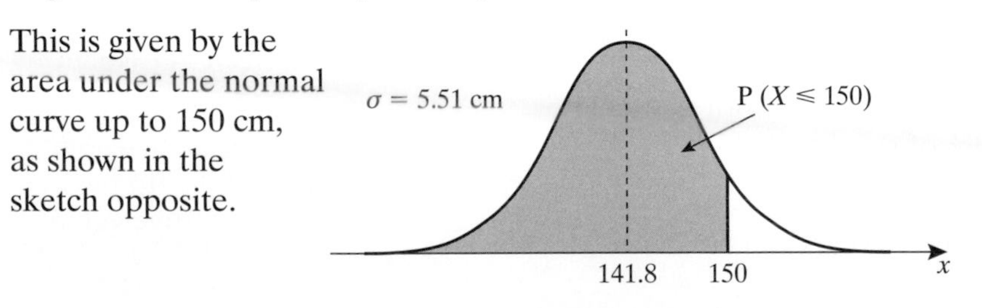

It is helpful to put all of the information from the situation onto a sketch.

You can use the *Standard Normal Table* introduced in **Section 3.2** to find this area if you standardise the variable. This means transforming the variable by subtracting the mean and then dividing by the standard deviation. All normal distributions are similar in shape and this transformation always gives the **standardised** normal distribution.

In this case:

$$P(X \leqslant 150) = P\left(\frac{X-\mu}{\sigma} \leqslant \frac{150-141.8}{5.51}\right)$$
$$= P(Z \leqslant 1.49) = \Phi(1.49)$$

Using the *Standard Normal Table* gives the result 0.9319.

This means that in theory about 93% of 10-year-old boys would be allowed to use the ride.

Standardising

If X is a normal variable with mean μ and standard deviation σ, then

$$Z = \frac{X-\mu}{\sigma}$$

is the standardised normal variable with mean 0 and standard deviation 1.

Note that, for a particular value X, the corresponding value Z is the number of standard deviations between X and the mean. Here 1.49 indicates that 150 cm is 1.49 standard deviations above the mean.

Activity 3.3A

1 Estimate the percentage of 10-year-old girls that will be allowed on the Octopus.

Now consider the minimum height restriction of 1.1 m on Calgary Stampede.

The probability that a 10-year-old boy taken at random will be able to go on the Calgary Stampede is $P(X \geqslant 110)$.

This is given by the area under the normal curve above 110 cm. The sketch below shows this area. It is often useful to show the relevant values of both X and Z on the same sketch.

Discussion point

Can you explain why Z gives the number of standard deviation between X and the mean?

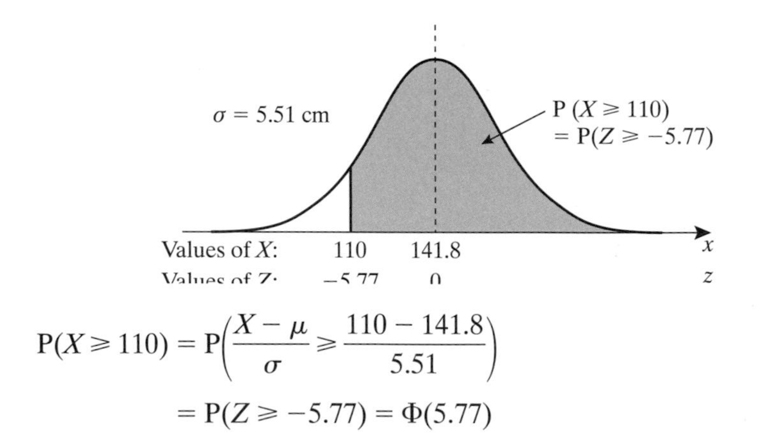

$$P(X \geqslant 110) = P\left(\frac{X-\mu}{\sigma} \geqslant \frac{110-141.8}{5.51}\right)$$
$$= P(Z \geqslant -5.77) = \Phi(5.77)$$

The standard normal table only goes as far as 3.79 with a value of 0.9999 corresponding to almost 100% of the distribution. For 5.77 the proportion would be even nearer 100%.

This means that virtually all boys of age 10 years will be allowed to use the ride.

Again note that all of the useful information has been written onto the sketch.

No attempt has been made to scale the sketch – the main thing is to show the mean, the value that is under consideration and the area that will give the probability.

You can see from the sketch that the shaded area is the large part – this is given from the table by looking up the corresponding positive value of Z.

Activity 3.3A (continued)

2 Estimate the percentage of 10-year-old girls that will be allowed on the Calgary Stampede.

If possible, share the rest of the work with other students and pool your results.

3 Estimate the percentage of girls and boys in other age groups that will be allowed on each of the rides.

4 Write a brief summary for the theme park manager describing how the height restrictions will affect the use of the rides by children aged between 5 and 15 years of age.

Discussion point

It is not really necessary to work out the percentages for every age.

Which calculations do you think should be done?

A variety of normal models

The following activity includes a variety of contexts in which you can use distributions to make predictions. Work through at least four of these and also **question 7**.

Excel Activity

Activity 3.3B

1 **Car speeds**

The speeds of cars passing a particular point on a road with a 30 mph speed limit are measured. The results suggest that the speeds are normally distributed with mean 34 mph and standard deviation 4 mph.

Estimate the percentage of cars that are

a breaking the speed limit

b travelling at more than 40 mph

c travelling at less than 25 mph.

2 **Video tapes**

A manufacturer makes 1, 2, 3 and 4 hour video tapes. Samples from the production line indicate that the actual tape lengths are normally distributed with the means and standard deviations given below.

Stated length	Mean (min)	SD (min)
1 hour	62	2
2 hours	125	4
3 hours	190	6
4 hours	250	6

For each type of tape, find the probability that a tape taken at random will be shorter than the stated length.

3 Intelligence quotient

Intelligence (IQ) scores are in theory normally distributed with mean 100 and standard deviation 16. Estimate the percentage of the population who have IQs that are:

a greater than 140

b less than 70

c between 90 and 110.

4 Athletes

An athletics club keeps records of the performances of its best athletes. The table below gives information about four of these athletes and the club record in each event.

Athlete	Mean	Standard deviation	Record
Triple jumper (men's)	16.94 m	0.36 m	17.51 m
Javelin thrower (women's)	57.65 m	2.75 m	63.18 m
100 m sprinter (women's)	10.89 s	0.24 s	10.92 s
1 mile runner (men's)	3 min 51.73 s	2.83 s	3 min 50.12 s

Assuming that these distributions are normal, find the probability in each case that the athlete will break the club record in the next attempt.

Discussion point

Is it fair to use a probability model in this way when analysing the performance of an athlete?

5 Exam marks

The percentage marks achieved by candidates in an examination can be modelled by a normal distribution with mean 58% and standard deviation 12%.

The table below shows the percentage marks that are awarded each of the pass grades A, B and C.

Grade	Percentage
A	75 and over
B	60–74
C	40–59
Fail	Less than 40

Use the normal model to predict the proportion of candidates that will achieve each grade.

Discussion point

What will you use as the boundaries here? For example, a mark of 75 and above achieves a grade A, a mark of 74 achieves a grade B.

6 High blood pressure

High blood pressure is one of our most chronic illnesses. Blood pressure is measured using a sphygmomanometer. An inflatable cuff is put around a person's arm and inflated until the blood flow is cut off. When air is let out of the cuff two measurements of the blood pressure are taken. The first of these values, called the systolic pressure, is the pressure in mmHg (millimetres of mercury) when the heart is contracting and blood starts to flow again. The second value, called the diastolic blood pressure, is the pressure when the heart is relaxing and blood flows continuously.

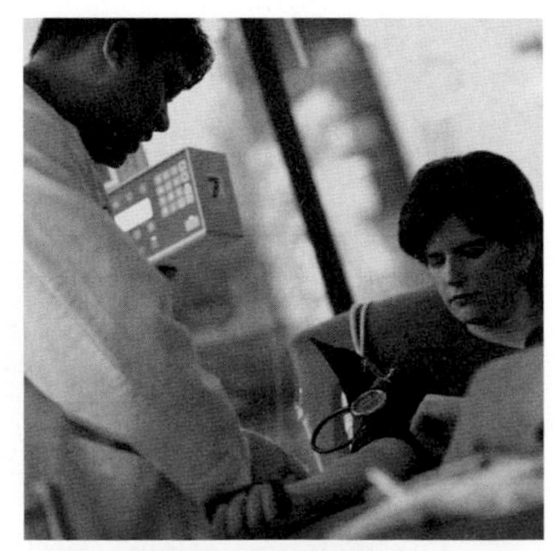

Hypertension is the medical term for high blood pressure

Surveys carried out by the Department of Health gave the following results for the adult population:

		Men	Women
Systolic blood pressure (mmHg)	Mean	137.3	133.2
	SD	19.7	22.5
Diastolic blood pressure (mmHg)	Mean	76.3	72.4
	SD	12.4	12.8

High blood pressure can be divided into the following categories:

Level of severity of hypertension	Mild	Moderate	Severe
Systolic blood pressure (mmHg)	140–160	160–200	Above 200
Diastolic blood pressure (mmHg)	90–105	105–115	Above 115

Estimate the number of people from a random sample of 1000 men and 1000 women who have blood pressures in each of these categories.

7 Features of normal distributions

Show that for any normal distribution with mean μ and standard deviation σ

a approximately two-thirds of the population lies between the values $\mu - \sigma$ and $\mu + \sigma$

b approximately 95% of the population lies between the values $\mu - 2\sigma$ and $\mu + 2\sigma$

c approximately 99.9% of the population lies between the values $\mu - 3\sigma$ and $\mu + 3\sigma$.

Superjumbo

The A380 is the world's first twin-deck, twin-aisle airliner. A typical upper deck layout provides 96 business and 103 economy class seats. The main deck provides 22 first class seats and 334 economy class seats.

When designing new aircraft, one major consideration is weight. Designers strive to minimise the weight of the aircraft. They also analyse the weight distribution when it is both empty and fully loaded. Although the weight of the passengers who will travel on each journey is unknown, it is possible to make predictions based on the mean and standard deviation of the weight distributions of the population.

Suppose you wish to estimate how many people in each section of the aircraft will have weights between 70 kg and 90 kg.

Estimates of the mean and standard deviation of the weights of adult males and females are given in the table below.

	Mean (kg)	Standard deviation (kg)
Males	80.6	14.7
Females	67.8	15.8

Denoting the height of a randomly chosen male by X,

$$P(70 \leqslant X \leqslant 90) = P\left(\frac{70-80.6}{14.7} \leqslant \frac{X-\mu}{\sigma} \leqslant \frac{90-80.6}{14.7}\right)$$

$$= P(-0.72 \leqslant Z \leqslant 0.64)$$

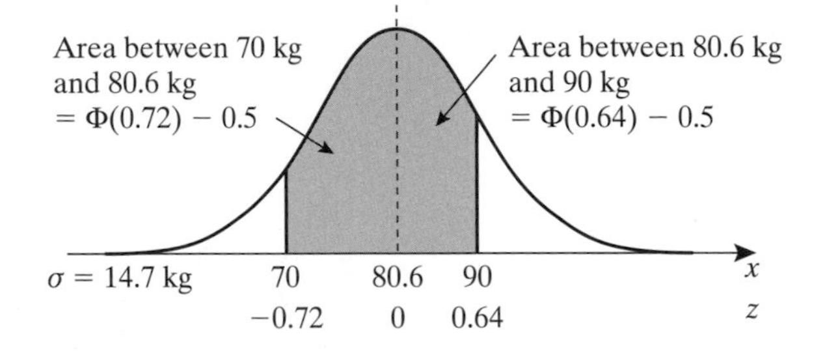

This area can be found in two parts, using the fact that the area at each side of the mean is 0.5:

$P(0 \leqslant Z \leqslant 0.64) \quad = \Phi(0.64) - 0.5 = 0.7389 - 0.5 = 0.2389$
$P(-0.72 \leqslant Z \leqslant 0) = \Phi(0.72) - 0.5 = 0.7642 - 0.5 = 0.2642$

Adding these gives:
$P(-0.72 \leqslant Z \leqslant 0.64) = 0.2389 + 0.2642 = 0.5031$

This probability is an estimate of the proportion of adult males who weigh between 70 kg and 90 kg.

If we assume that the seats in each section of the aircraft are equally divided between male and females, then the number of males in each section would be:

Upper deck 48 business class and 52 economy class
Main deck 11 first class and 167 economy class

The expected number of males weighing between 70 kg and 90 kg in each section of the aircraft is found by multiplying the probability by the total number of males in each section.

Upper deck	Business Class	0.5013×48	= 24 males
	Economy Class	0.5013×52	= 26 males
Main Deck	First Class	0.5013×11	= 6 males
	Economy Class	0.5013×167	= 84 males

☞ **Probability** is an estimate of the **proportion**.

Expected frequency = probability × total frequency

It is possible to use linear interpolation to give more accurate results from the table. Using 3 significant figures here gives $P(-0.721 \leqslant Z \leqslant 0.639)$.

From the table
$\Phi(0.72) = 0.7642$ and
$\Phi(0.73) = 0.7673$

Using linear interpolation
$\Phi(0.721) \approx 0.7642 + \frac{1}{10} \times 0.0031$
$\Phi(0.721) \approx 0.7645$

Similarly using
$\Phi(0.63) = 0.7357$ and
$\Phi(0.64) = 0.7389$ gives
$\Phi(0.639) \approx 0.7357 + \frac{9}{10} \times 0.0032$
$= 0.7386$

The result is then
$P(-0.721 \leqslant Z \leqslant 0.639)$
$= 0.2645 + 0.2386 = 0.5031$

Note that although individual results such at $\Phi(0.721)$ and $\Phi(0.639)$ are more accurate when their difference is found in this case you get the same final answer.

Discussion points

Do you think this assumption is realistic?

How would you estimate the number of men in each section?

Activity 3.3C

1 Assuming that the rest of the seats are occupied by adult females, estimate the number of females weighing between 70 kg and 90 kg in each section of the aircraft. Hence estimate the total number of passengers weighing between 70 kg and 90 kg in each section of the aircraft.

2 Estimate the total number of passengers weighing

a between 90 kg and 110 kg

b more than 110 kg

in each section of the aircraft.

Practice Sheet: Normal Distributions

Fits like a glove

A manufacturer is planning a new range of gloves for adults. There will be three sizes for women: small, medium and large. There will be five sizes for men: extra small, small, medium, large and extra large. Each of the women's sizes is to cater for roughly a third of the adult female population and each of the men's sizes for a fifth of the adult male population. Measurements from a large sample of the adult population have given the following results.

		Males	Females
Hand length (mm)	Mean	189.8	175.0
	SD	9.8	9.4
Hand width (mm)	Mean	91.6	78.3
	SD	5.9	4.7

The designer can find the range of hand sizes that are to be catered for by each glove size by assuming that each of these distributions is normal.

Consider the distribution of female hand lengths.

The sketch shows the distribution split into three sections, each containing a third of the female population. As the total area under the curve is 1, the area of each section is $0.\dot{3}$.

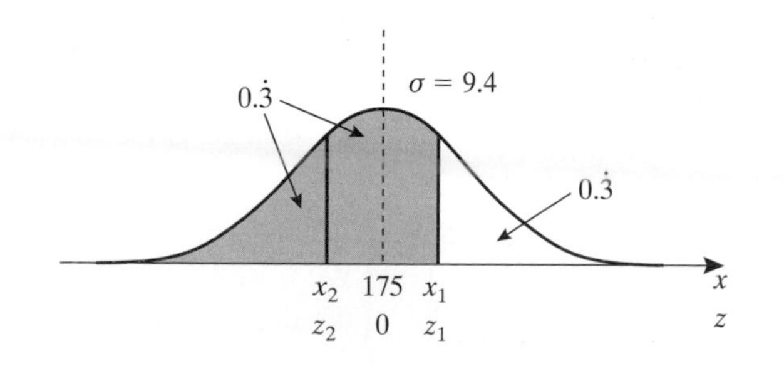

Discussion points

Do you think this is a good way of deciding the boundaries for the three sizes?

What other ways are there of doing this?

In this case the areas are known and the values of X at the boundaries of each section are required.

The area below the boundary value z_1 is $\frac{2}{3}$ i.e. $0.\dot{6}$.

This gives $\Phi(z_1) = 0.\dot{6}$.
Look in the central part of the *Standard Normal Table*. The value nearest to $0.\dot{6}$ is 0.6664 corresponding to a Z value of 0.43.

Using $z_1 = \dfrac{x_1 - \mu}{\sigma}$ gives $0.43 = \dfrac{x_1 - 175}{9.4}$
Rearranging gives $x_1 = 175 + 0.43 \times 9.4 = 179.042$
This seems a reasonable answer when you compare it with the sketch diagram.

By symmetry the other boundary of Z, z_2, is -0.43.

Using $z_2 = -0.43$ gives $x_2 = 175 - 0.43 \times 9.4 = 170.958$.
Again this seems a reasonable answer when you compare it with the sketch diagram.
So the boundaries for hand lengths for the female gloves are 171 mm and 179 mm (to 3 significant figures).

To find a value of X

1 Find the value of Z by using the *Standard Normal Table* in reverse.

2 Use the relationship $Z = \dfrac{X - \mu}{\sigma}$ to find the corresponding value of X.
Rearranging gives an expression for X in terms of Z:
$X = \mu + Z\sigma$

Activity 3.3D

1 Find the width boundaries for each size of women's gloves. Assume the sizes will each cater for approximately one-third of the population.

In reality, the designer also needs to know the minimum length for the small size and the maximum length for the large size. These depend on how sure the manufacturer wants to be of including all of the population.

Suppose the target market is the 95% of the female population whose hand lengths lie within two standard deviations of the mean. This gives a lower boundary of $175 - 2 \times 9.4 = 156.2$ mm and an upper boundary of $175 + 2 \times 9.4 = 193.8$ mm.

To the nearest millimetre, the designer should aim to design a small size catering for hand lengths from 156 mm to 171 mm, a medium size catering for hand lengths from 171 mm to 179 mm and a large size catering for hand lengths from 179 mm to 194 mm.

Discussion points

The small and large sizes will not now be designed for *exactly* a third of the population. What percentage of the population will they now be designed to suit?

What difference would it make if the minimum and maximum hand lengths were taken to be three standard deviations from the mean? What advantages and disadvantages would this have?

The manufacturer could have decided the length boundaries for each size by simply splitting the length range between $\mu \pm 2\sigma$ into three equal sections. What would the boundaries have been? What proportion of the female population would be catered for by each size? What are the advantages and disadvantages of this method?

Activity 3.3D (continued)

2 Boundaries for the men's sizes of extra small, small, medium, large and extra large are to be chosen so that each size caters for about a fifth of the population.
Find the range of hand *lengths* that are to be catered for by each size of men's gloves.

3 Find the range of hand ***widths*** that are to be catered for by each size of men's gloves. State any assumptions you make.

Computer workstation

The computer workstation problem in **Section 3.1** concerned an adjustable monitor screen. The minimum height for the top of the monitor screen was taken at the 5th percentile of the distribution of female eye heights and the maximum height at the 95th percentile of the distribution of male eye heights.

Activity 3.3E

Use the results in the table to find the 5th percentile of female eye heights and the 95th percentile of male eye heights.

	Sitting eye height (mm)	
	Mean	Standard deviation
Female	1137	49
Male	1236	53

Compare your results with those from **Activity 3.1B**.

Using tables in reverse

Activity 3.3F gives a variety of problems in which it is necessary to use the *Standard Normal Table* in reverse. Work through at least two of the problems.

Activity 3.3F

1 **Journey time**
A motorist records the time taken for his journey to work each morning. He finds that the journey time is normally distributed with mean 45 minutes and standard deviation 5.2 minutes. Find how long the motorist should allow for the journey if he wants to arrive at work on time on

a 90% of workdays **b** 95% of workdays

c 99% of workdays.

2 **Growth**
Assume that the distributions of weights of boys and girls at different ages is normal with the means and standard deviations given in the table below.

	Boys		Girls	
Age	Mean (kg)	SD (kg)	Mean (kg)	SD (kg)
4	18.0	2.1	17.5	2.4
8	27.0	4.3	27.0	4.7
12	40.5	7.9	42.0	8.3
16	61.5	9.3	55.5	7.5

Take 'on time' to include being early for work.

Growth charts

Clinical growth charts are an important tool for monitoring children's development. They usually show the median, quartiles and other percentiles of the weight and height of children of different ages. Some growth charts can be found on the Internet.

If possible, download some growth charts and compare what they show with your answers.

a For each gender and age estimate

i the upper quartile **ii** the lower quartile
iii the 10th percentile **iv** the 90th percentile
v the 3rd percentile **vi** the 97th percentile.

b Write a brief summary of what the results show about the way in which the weights of children vary at different ages.

3 Guarantees

A manufacturer of light bulbs wants to offer guarantees on the lifetimes of some of its products. Samples of the products suggest the lifetimes follow normal distributions with the means and standard deviations given below.

Type	Mean life	Standard deviation
Standard	1259 hours	82.4 hours
Low energy fluorescent	11 325 hours	217.3 hours
Fluorescent tube	9653 hours	384.5 hours

The manufacturer wants to give refunds on no more than 2% of each product. For each type of light bulb find, to the nearest 100 hours, the maximum lifetime that should be guaranteed.

4 Jars of jam

The labels on jars of jam state that they contain 454 grams. When a sample of 500 jars are tested it is found that the jam in 88 of the jars weighs less than 454 grams and the jam in 36 of the jars weighs more than 460 grams.

Assume that the mass of jam in a jar is normally distributed with mean μ grams and standard deviation σ grams.

a Use the *Standard Normal Table* and the results from the sample to find two equations relating μ and σ.

b Solve the equations to find values for μ and σ. Give your answers to 2 decimal places.

The jar-filling process is altered so that the mean mass of jam in the jars is 458 grams. The standard deviation is not altered.

c Estimate the percentage of jars that will now contain less than 454 grams of jam.

College journey times

In a questionnaire students were asked how long, to the nearest minute, it had taken them to travel to college.

The table gives the results.

In **Activity 3.3G** you will investigate how closely a normal distribution models this distribution of times.

Time (minutes)	No. of students
0–9	35
10–14	732
15–19	954
20–24	1746
25–29	2502
30–34	2162
35–39	1895
40–49	1254
50–59	713
60 or over	27

Activity 3.3G

1 Draw a histogram to illustrate the results in the table.

2 Calculate estimates of the mean and standard deviation of the journey times.

3 Use your histogram to estimate the proportion of the times that are

 i within one standard deviation of the mean
 ii within two standard deviations of the mean
 iii within three standard deviations of the mean.

 Briefly comment on how your answers compare with what you would expect from a normal distribution.

4 Assuming a normal distribution with the mean and standard deviation you have found in **part a**, calculate the expected number of students in each group.

5 i Use the expected frequencies found in **question 4** to draw a histogram.
 ii Compare the histogram of expected frequencies with the histogram of the real data. Describe the similarities and differences.

The process of modelling a distribution by a normal distribution in this way is often called 'fitting a normal distribution'.

3.4 Revision summary

Describing distributions

When a distribution is not symmetrical it is said to be **skewed**. The sketches show a **positively skewed** distribution in which the mean is greater than the median and a **negatively skewed** distribution in which the mean is less than the median.

A **bi-modal** distribution has two peaks. A distribution with more than two peaks is said to be **multi-modal**.

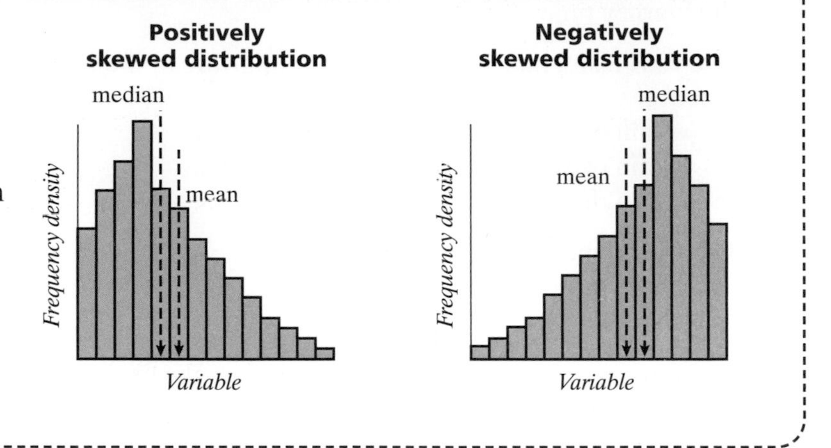

Relative frequency density histograms

$$\text{Frequency density} = \frac{\text{frequency}}{\text{class width}} \qquad \text{Relative frequency density} = \frac{\text{frequency density}}{\text{total frequency}}$$

The area of a bar on a histogram in which frequency density is plotted against a continuous variable gives the **frequency** of the group which the bar represents.

The area of a bar on a relative frequency density histogram gives the **proportion** of the sample in that group. The total area under a relative frequency density histogram is 1.

Features of a normal distribution

A normal distribution is a theoretical model. The histogram is bell-shaped and perfectly symmetrical about its central value.

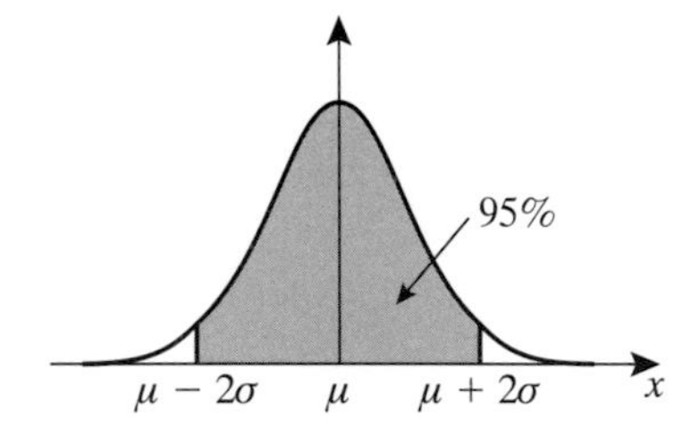

The mean, mode and median of a normal distribution are equal and lie at the line of symmetry.

In a normal distribution:

- about two-thirds of the distribution lies within one standard deviation of the mean
- about 95% of the distribution lies within two standard deviations of the mean
- about 99% of the distribution lies within three standard deviations of the mean.

Standard normal distribution

The standard normal variable, Z, has mean 0 and standard deviation 1.

The standard normal distribution

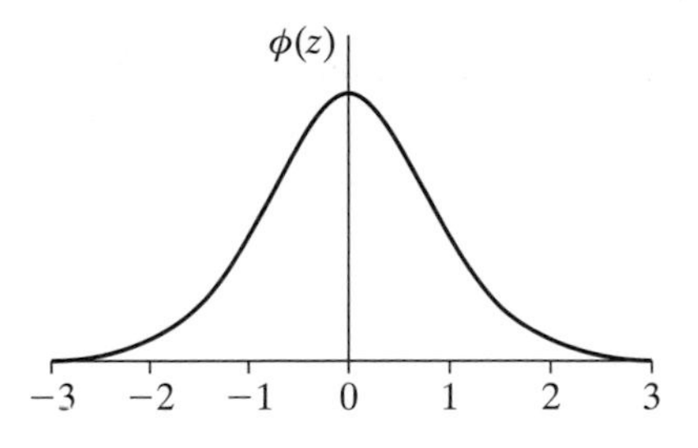

The **area under the graph** gives **probability.**

Area values can be looked up in a *Standard Normal Table*. Probabilities for negative values of Z can be found using symmetry.

Standardising other normal distributions

If X is a normal variable with mean μ and standard deviation σ, the standardised normal variable is given by $Z = \dfrac{X - \mu}{\sigma}$

Rearranging gives $X = \mu + Z\sigma$

For a particular value x, the corresponding value of z is the number of standard deviations between x and the mean.

To solve problems, draw a sketch showing all the information, shade relevant areas, then use the *Standard Normal Table*.

Expected frequency
= probability × total frequency

3.5 Preparing for assessment

Your coursework portfolio

The work you have just completed in this chapter will allow you to consider in detail data that can be modelled using the normal distribution.

You may well be able to model naturally occurring data you have collected as part of your work for another subject using the normal distribution. You may be able to see this by drawing a stem and leaf diagram or a histogram. If you suspect this is the case you can check by finding the proportion of the data that lies within one or two standard deviations of the mean – for a normal distribution, this is approximately 68% and 95% respectively. If your data can indeed be modelled by the normal distribution then you can use the ideas you have learnt in this chapter to solve problems and make predictions about the situation.

Note: The Nuffield FSMQ website provides many datasets and other resources that may be of use to you.

Practice exam questions

Data

1 BSI 1990 (UK) was a large-scale anthropometric survey of children carried out in the UK. The table gives the results obtained by measuring the head circumferences of Caucasian children from mainly urban areas.

Age (years)	Sex	Mean (cm)	Standard deviation (cm)
5	Male	51.74	1.46
	Female	50.82	1.69
6	Male	52.05	1.56
	Female	51.64	1.39
7	Male	52.32	1.46
	Female	52.35	1.44
8	Male	52.76	1.47
	Female	52.43	1.68
9	Male	53.06	1.82
	Female	52.67	1.59
10	Male	53.56	1.53
	Female	53.35	1.53

Question

Assume that for each age and gender of child, head circumference follows a normal distribution with the mean and standard deviation given in the table.

a Use the statistical measures in the table to compare and contrast the head circumferences of 9-year-old boys with those of 9-year-old girls.

b State an estimate of

i the mode

ii the median

head circumference of 9-year-old boys.

c A clothing manufacturer makes baseball caps that will fit heads with circumferences from 51 centimetres to 54 centimetres. Find an estimate for the percentage of 9-year-old boys that such a baseball cap will fit.

d A statistician says that the data may not give accurate predictions because the samples used were not representative. Describe two ways in which the samples could be more representative of the UK population.

Data

2 The sketch shows a person standing erect with an arm stretched vertically upwards as if reaching for something from a shelf.

Overhead reach is the distance measured vertically from the floor to the middle of the pad of the person's thumb where contact is made in a pinch grip.

The table below gives the means and standard deviations of overhead reach measured from samples of male and female adults in the UK.

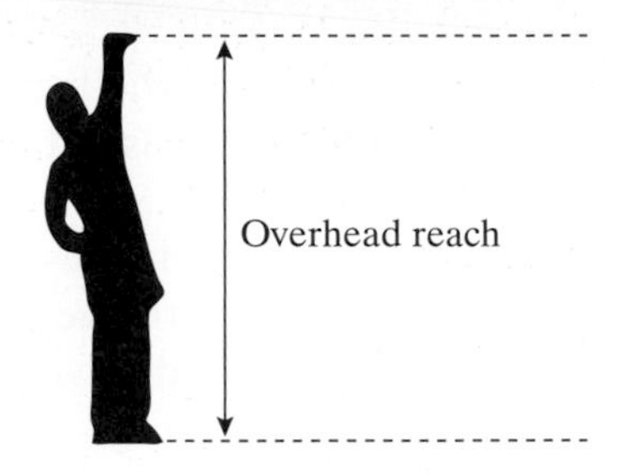

	Mean (mm)	Standard deviation (mm)
Male	2125.3	84.5
Female	1987.3	79.3

Question

Assume both distributions of overhead reach are normal.

a **i** Draw a sketch of the distribution of female overhead reach with the mean value marked.

ii On your sketch shade the part of the distribution where the female overhead reach is greater than the mean male overhead reach.

iii Calculate an estimate of the percentage of females who have an overhead reach greater than the mean male overhead reach. Show all your working.

b A bookshelf designer wants to accommodate overhead reaches from the 5th percentile in the distribution of female overhead reaches to the 95th percentile in the distribution of male overhead reaches.

Estimate

i the 95th percentile in the distribution of male overhead reach

ii the 5th percentile in the distribution of female overhead reach.

Data

3 Eggs are sold in four different sizes.

Size	Weight (g)
Very large	73 and over
Large	63–73
Medium	53–63
Small	53 and under

Question

A poultry farmer weighs each egg in a large sample from his hens.

a **i** Describe two features that you would expect a histogram of these measurements to have if the weights follow a normal distribution.

ii Describe two calculations the farmer could carry out to test whether the distribution is normal.

b The sample of eggs has a mean weight of 65.4 grams and standard deviation 5.9 grams. Assuming that the weights of the eggs laid by the hens are normally distributed, estimate the percentage of eggs laid by the flock that can be expected to be in each of the different sizes.

c The farmer expects his hens to lay 2000 eggs per week. He decides to keep the small eggs to distribute to his family and friends and sell the rest. Estimate how many small eggs there will be per week.

4 Interpreting Tables and Diagrams

Contents

A study investigated whether driving tests were consistent or not. Candidates who had already booked a driving test at a number of test centres were asked to take a second driving test, free of charge, within a few days of the first. Participants were not told the result of the first test, or given any feedback on how they had done, until after they had completed the second. At the end of the second test they were issued with a pass certificate if they had passed either or both the tests. The second examiner was not told the result of the first test until after the second test had finished. Test routes and times were allowed to vary as would happen normally with test allocation, to ensure that these sources of variability were included. The table shows the main results. What can you conclude?

		Results of 1st test		
		Fail	Pass	Total
Results of 2nd test	Fail	42%	16%	58%
	Pass	20%	22%	42%
	Total	63%	37%	100%

Quality control charts are used to monitor quality in a production process. Samples are taken regularly from the production line and the mean measurement from each sample is plotted on a control chart.

The chart has a central line showing the theoretical mean. Other horizontal lines give warnings when the production process may be going out of control and indicate when action should be taken to rectify the problem.

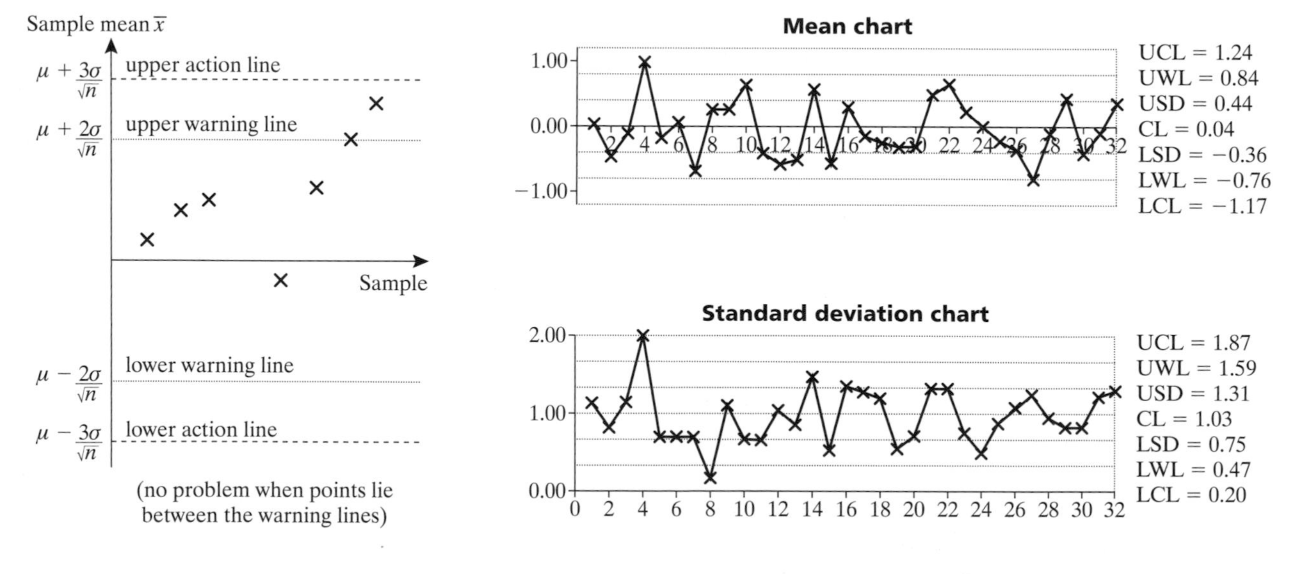

Already in this book you have met a lot of statistical data presented in a wide range of formats. In many cases either primary or secondary data is presented in tables or diagrams. You will have the opportunity to examine a range of statistical tables and diagrams and interpret what they tell you about the real situation they represent when you work through this chapter.

4.1 Tables of data

Drug trafficking

Drug trafficking
Breath tests
Population
House prices

The table on the next page gives the number of drug trafficking offences in the member states of the European Union for the years from 1995 to 1999 inclusive. Use the table to complete the following activity.

Activity 4.1A

1 Name the country that had

a the lowest number of drug trafficking offences in
- **i** 1995
- **ii** 1999

b the highest number of drug trafficking offences in
- **i** 1995
- **ii** 1999

c the greatest percentage increase in drug trafficking offences between 1995 and 1999

d the greatest percentage decrease in drug trafficking offences between 1995 and 1999.

2 **a** Write down calculations to show that the number of drug trafficking offences in the USA
- **i** increased by 62% between 1995 and 1999
- **ii** increased by 8% between 1998 and 1999.

b Write down calculations to show that the number of drug trafficking offences in England and Wales
- **i** fell by 6% between 1995 and 1999
- **ii** fell by 8% between 1998 and 1999.

3 **a** In which countries did the number of drug trafficking offences more than double between 1995 and 1999?

b In which countries did the number of drug trafficking offences decrease by more than a quarter between 1995 and 1999?

4 The number of drug trafficking offences in England and Wales decreased by 8% between 1998 and 1999. If the number increased by 8% between 1999 and 2000, would the number of drug trafficking offences return to the same number as in 1998? Explain your answer using calculations.

$$\text{Percentage change} = \frac{\text{final value} - \text{original value}}{\text{original value}} \times 100$$

5 Either:

a Write a paragraph comparing the number of drug trafficking offences in England and Wales, Northern Ireland and Scotland and the way in which they varied between 1995 and 1999.

Or:

b Write a brief article for a newspaper that highlights your main observations from this table.

Crimes[1] recorded by the police: Drug trafficking[2]

Country	1995	1996	1997	1998	1999	% change 1995–99	% change 1998–99
EU Member States average						***31%***	***3%***
England & Wales[3]	21 272	22 122	23 336	21 788	19 956	−6%	−8%
Northern Ireland[3]	358	204	176	193	193	..	0%
Scotland	7974	6957	8180	8490	8668	9%	2%
Austria	11 635	14 923	16 808	15 906	16 324	40%	3%
Belgium	6896	8362	11 072	10 158	9968	45%	−2%
Cyprus[4]	156	183	239	285	252	62%	−12%
Czech Republic[5]	309	608	789	1029	1576	410%	53%
Denmark	291	354	171	178	129	−56%	−28%
Estonia[6]	12	30	30	117	180	1400%	54%
Finland[4]	9052	7868	8323	9461	11 674	29%	23%
France	79 052	79 617	86 961	92 858	101 841	29%	10%
Germany	54 889	65 615	69 093	68 994	73 271	33%	6%
Greece[4]	2930	4272	5970	6574	6692	128%	2%
Hungary	429	440	943	2068	2860	567%	38%
Ireland (Eire)	826	958	1637	1766	1971	139%	12%
Italy	38 269	38 954	41 420	43 014	45 038	18%	5%
Luxembourg[4]	764	864	805	825	941	23%	14%
Netherlands	3473	6593	10 300	7700	7600	119%	−1%
Norway[4]	23 567	27 657	34 705	38 990	41 229	75%	6%
Poland	284	494	994	2043	2063	626%	1%
Portugal	4509	3879	3390	3538	4091	−9%	16%
Russia[4]	79 819	96 645	185 832	190 127	216 364	171%	14%
Slovenia	454	680	972	997	1121	147%	12%
Spain	15 118	15 307	14 274	13 263	..	−12%[11]	−7%[12]
Sweden[7]	689	635	561	446	471	−32%	6%
Switzerland	2171	2515	3253	3734	3715	71%	−1%
Australia[8]	..	24 994	24 313	23 348	17 702	..	−24%
Canada	17 394	17 913	17 299	17 808	19 966	15%	12%
Japan[9]	2982	2678	2359	2712	2299	−23%	−15%
New Zealand[4]	12 274	12 658	14 532	15 158	..	23%[11]	4%[12]
South Africa[4]	40 782	39 241	42 805	39 830	41 461	2%	4%
U.S.A.[10]	24 915	27 457	33 160	37 322	40 383	62%	8%

(1) Definitions of offences vary between countries both due to legal differences and statistical recording methods; comparisons may be affected by these differences.
(2) Illegal importing, exporting, supplying, transportation, etc. of narcotic drugs.
(3) By financial year from 1997 (e.g. 1997 = 1 April 1997 to 31 March 1998).
(4) All drugs offences.
(5) Number of people prosecuted.
(6) Illegal manufacture, acquisition, storage, transportation, delivery or trafficking of narcotic drugs or psychotropic substances.
(7) Includes attempts, preparation and conspiracy to commit an offence.
(8) By financial year (e.g. 1994 = 1 July 1993 to 30 June 1994).
(9) Excluding amphetamines.
(10) Arrests by Drug Enforcement Administration.
(11) 1995–1998. (12) 1997–1998. .. Data not available.

Breath tests

When data or information is given as percentages it is important to read the information carefully to understand what the percentages are and how they have been calculated.

The table below gives data about breath tests carried out in Great Britain between 1995 and 1999.

Road accidents: breath tests performed on car drivers and motorcycle riders involved in injury accidents: Great Britain: 1995–1999

	Number/*percentage*				
	1995	**1996**	**1997**	**1998**	**1999**
Car drivers involved in injury accidents	318 083	331 091	338 924	337 794	329 866
Breath tested[1]					
Number	99 631	133 347	157 373	173 610	175 916
Percentage of drivers involved in injury accidents	31	40	46	51	53
Failed breath test[1]					
Number	6639	7303	7087	6690	6669
Percentage of drivers tested	7	5	5	4	4
Motorcycle riders involved	24 219	23 798	25 211	25 514	27 122
Breath tested[1]					
Number	5720	7906	9926	11 416	12 970
Percentage of riders involved	24	33	39	45	48
Failed breath test[1]					
Number	438	408	428	426	443
Percentage of riders tested	8	5	4	4	3

[1] Includes refusals.

Discussion point

Choose some values from the table at random. Can you tell at a glance what information each value gives?

The first percentage in each column is the **percentage of car drivers involved in injury accidents who were breath tested.**

The calculation giving this percentage for 1999 is:

$$\frac{175\,916}{329\,866} \times 100 = 53.33 = 53\% \text{ to the nearest per cent.}$$

Note how the numerators and denominators change depending on what percentages are being calculated.

The second percentage in each column is the **percentage of car drivers in injury accidents asked for a breath test who failed (or refused).**

The calculation giving this percentage for 1999 is:

$$\frac{6669}{175\,916} \times 100 = 3.79 = 4\% \text{ to the nearest per cent.}$$

Activity 4.1B

1 In 1999, 6669 drivers failed or refused a breath test out of the 329 866 drivers involved in injury accidents.

  a Use these data to find the percentage of all car drivers involved in injury accidents in 1999 who failed (or refused) a breath test.

  b You can find the answer to **a** by finding 3.79% of 53.33%. Explain why.

2 **a** Show how the last two percentages in the 1999 column have been calculated.

  b Use two different methods to find the percentage of *all* the motorcycle riders involved in injury accidents in 1999 who failed (or refused) a breath test.

3 **a** Find the percentage increase between 1995 and 1999 in
  - i the number of car drivers involved in injury accidents who failed (or refused) breath tests
  - ii the number of motorcycle riders involved in injury accidents who failed breath tests
  - iii the total number of car drivers and motorcycle riders involved in injury accidents who failed (or refused) breath tests.

  b Briefly interpret your answers to part **a**.

4 **a** What percentage of the total number of car drivers and motorcyclists involved in injury accidents in 1997 were
  - i car drivers
  - ii motorcyclists?

  b What percentage of the total number of failed breath tests shown in the 1997 column of the table were taken by
  - i car drivers involved in injury accidents
  - ii motorcycle riders involved in injury accidents?

  c Briefly interpret your answers to **parts a** and **b**.

5 **a** What was the percentage increase between 1995 and 1999 in the number of
  - i car drivers involved in injury accidents
  - ii motorcycle riders involved in injury accidents?

  b Briefly interpret your answers to **part a**.

6 **a** What was the percentage change in the number of failed (or refused) breath tests taken by car drivers involved in injury accidents between 1996 and 1999?

  b What was the percentage increase in the number of failed (or refused) breath tests taken by motorcycle riders involved in injury accidents between 1996 and 1999?

Discussion point
How would you assess to what extent drinking is responsible for injuries in road accidents? What other information would be useful?

Discussion point
What other information would you collect if you were asked to write a report comparing how safe car drivers and motorcycle riders are on the roads?

Population

The table below shows the yearly percentage increase in population for each country in the UK.

Year	England	Wales	Scotland	N. Ireland	UK
1992	0.35	0.28	0.08	1.12	0.34
1993	0.32	0.24	0.18	0.80	0.32
1994	0.36	0.24	0.23	0.61	0.35
1995	0.40	0.14	0.10	0.42	0.36
1996	0.38	0.14	−0.18	0.85	0.33
1997	0.40	0.21	−0.10	0.66	0.35
1998	0.43	0.20	−0.06	0.54	0.38
1999	0.52	0.14	−0.02	0.18	0.45
2000	0.49	0.31	−0.08	0.35	0.43

This table was drawn up using the mid-year population figures given in the publication *'Population Trends Spring 2002'*. So, for example, the first entry, 0.35, indicates that the population of England increased by 0.35% between mid-1991 and mid-1992.

If the actual population in any of these years is known, then the population for other years can be calculated. For example, using the fact that the actual population of Scotland in 1994 was 5132 thousand, to the nearest thousand, you can estimate the population in 1995 by increasing it by 0.1%. The quickest way to do this is to multiply by 1.001.

Make sure you understand how to find and use these increase factors.

As a decimal 0.1% is 0.001.

$$P + 0.001P = (1 + 0.001)P = 1.001P$$

$$P - 0.0018P = (1 - 0.0018)P = 0.9982P$$

Population of Scotland in 1995 $\approx 1.001 \times 5\,132\,000 = 5\,137\,132$
$= 5137$ thousand

The population then falls by 0.18%.
Population of Scotland in 1996 $\approx 0.9982 \times 5\,137\,132 = 5\,127\,885$
$= 5128$ thousand

To find the population in previous years you need to divide by the increase factor instead of multiplying. For example, starting with the population of 5132 thousand in 1994:

Population of Scotland in 1993 $\approx \dfrac{5\,132\,000}{1.0023} = 5\,120\,224$

$= 5120$ thousand

This is because Population in 1994 = 1.0023 × Population in 1993

So that Population in 1993 = $\dfrac{\text{Population in 1994}}{1.0023}$

Activity 4.1C

1 Estimate the population of Scotland in

a 1997 **b** 1998 **c** 1999
d 2000 **e** 1992 **f** 1991

2 The population of England in 1995 was 48 903 thousand, to the nearest thousand. Estimate the population of England in

a 1996 **b** 1997 **c** 1998 **d** 1999
e 2000 **f** 1994 **g** 1993 **h** 1992

3 **a** Calculate $1.0028 \times 1.0024 \times 1.0024 \times 1.0014 \times 1.0014$

b Hence write down the percentage increase in the population of Wales between 1991 and 1996.

c Calculate the percentage increase in the population of Wales between 1996 and 2000.

4 The UK population in 2000 was approximately 59 756 thousand. Estimate the UK population in 1991.

House prices

The table below shows how the prices of dwellings (houses, flats and so on) in the UK varied between 1989 and 1999. The figures in the table are index numbers based on prices in the year 1993. Note that the indices in the 1993 row are all 100. All the other figures in the table give a comparison of the prices in other years with those in 1993. This therefore takes account of inflation in the value of money.

Index numbers are calculated in terms of the size of a variable during a **base year**. The size of the variable during the base year is usually taken to be 100.

Look at the column showing the prices of all dwellings in the UK. The first entry, 109.5, means that prices in the UK in 1989 were on average 9.5% higher than they were in 1993. Prices actually fell between 1989 and 1993. However, by 1997 prices had overtaken the 1989 values and by 1999 they had risen to 44.6% more than in 1993.

Dwelling price mix-adjusted indices[1][2]

All lenders 1993 = 100

	England	Wales	Scotland	Northern Ireland	United Kingdom		
					All	New	Second hand
1989	112.6	97.4	69.7	83.3	109.5	108.5	109.3
1990	110.6	101.5	75.9	88.1	108.1	107.5	108.2
1991	108.3	100.1	87.8	94.4	106.6	104.5	106.9
1992	103.3	99.0	93.2	96.1	102.6	101.6	102.8
1993	100.0	100.0	100.0	100.0	100.0	100.0	100.0
1994	102.7	101.3	101.1	103.9	102.5	100.1	102.9
1995	103.3	99.4	102.2	116.0	103.2	104.7	103.0
1996	106.9	103.8	105.3	126.0	106.9	109.7	106.5
1997	117.2	109.8	111.4	140.0	116.9	120.7	116.3
1998	130.6	115.0	117.7	154.9	129.7	128.3	129.9
1999	146.4	124.1	120.4	170.0	144.6	147.9	144.1

(1) Based on mortgages completed and adjusted for the mix of dwellings sold.
(2) Data up to and including 1992 was based on returns from Building Societies only. Data from 1993 onwards is based on returns from all mortgage lenders.

Activity 4.1D

1. A certain new house in the UK cost £67 500 in 1993. Use the UK New column to estimate the cost of a similar house in

   a 1999 **b** 1989.

2. If a certain new flat in the UK cost £40 000 in 1989, estimate the price of a similar flat in

   a 1993 **b** 1999.

3. Estimate the percentage increase in UK second-hand houses between

   a 1993 and 1996 **b** 1996 and 1999 **c** 1990 and 1993.

4. Compare and contrast the way house prices in England, Wales, Scotland and Northern Ireland varied between 1989 and 1999.

Discussion point

To estimate the flat's price in 1993 you need to calculate

$$\frac{100 \times £40\,000}{108.5}$$

Can you explain why?

Nuffield Resource
Skills activities:
'Indicators of child well-being'
'Athletics'
'Datasets'
'Trends'

4.2 Statistical charts and graphs

Subsidence
Population pyramids
Candlestick charts
Interpreting statistical diagrams

Subsidence

The chart below shows the total amount claimed from insurance companies due to subsidence in the UK between 1975 and 1997. Superimposed on the bar chart is a line graph giving the total rainfall for England and Wales from April to September (inclusive) during these years. The subsidence claim costs are in original-year values given in both pounds sterling and dollars using two scales at the left-hand side of the chart.

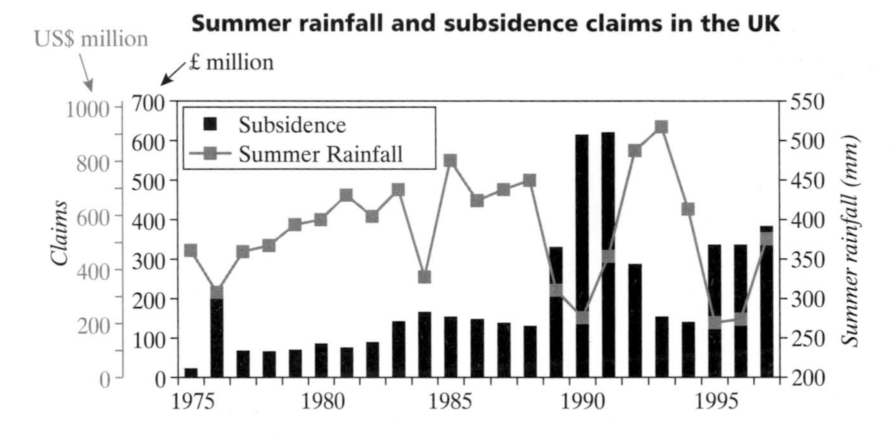

Land sinking or settling is called **subsidence**. This is often caused by underground mining and can cause damage to buildings.

Charts such as this, showing more than one quantity and using more than one scale, are used in many other real contexts.

Activity 4.2A

1 **a** In which year were subsidence claims greatest?

b What was the total amount claimed for subsidence during this year

i in £ **ii** in $?

c Estimate the range in the amounts claimed for subsidence during the period 1975–1997

i in £ **ii** in $.

d Estimate the percentage increase in the claims made for subsidence between 1988 and 1991.

e It is stated above that *'The subsidence claim costs are in original-year values …'*. Explain what this means and why the chart may be slightly misleading because of this.

2 **a** During which year was the summer rainfall

i greatest **ii** least?

b Estimate the mean rainfall per summer month in 1988.

c Between which consecutive years was the increase in summer rainfall greatest?

3 The Intergovernmental Panel on Climate Change (IPCC) states that there is *'a clear relationship between the cost of subsidence claims and rainfall (with some lag effects)'*. Explain what this means and identify any sections of the chart that support or contradict this statement.

Discussion point

Roughly what rate of exchange between pounds and dollars is used on this chart?

Population pyramids

Population pyramids are back-to-back bar charts showing the age distributions of the male and female populations of a country. Those below show the populations of the UK and India in 2000 and projections for 2050.

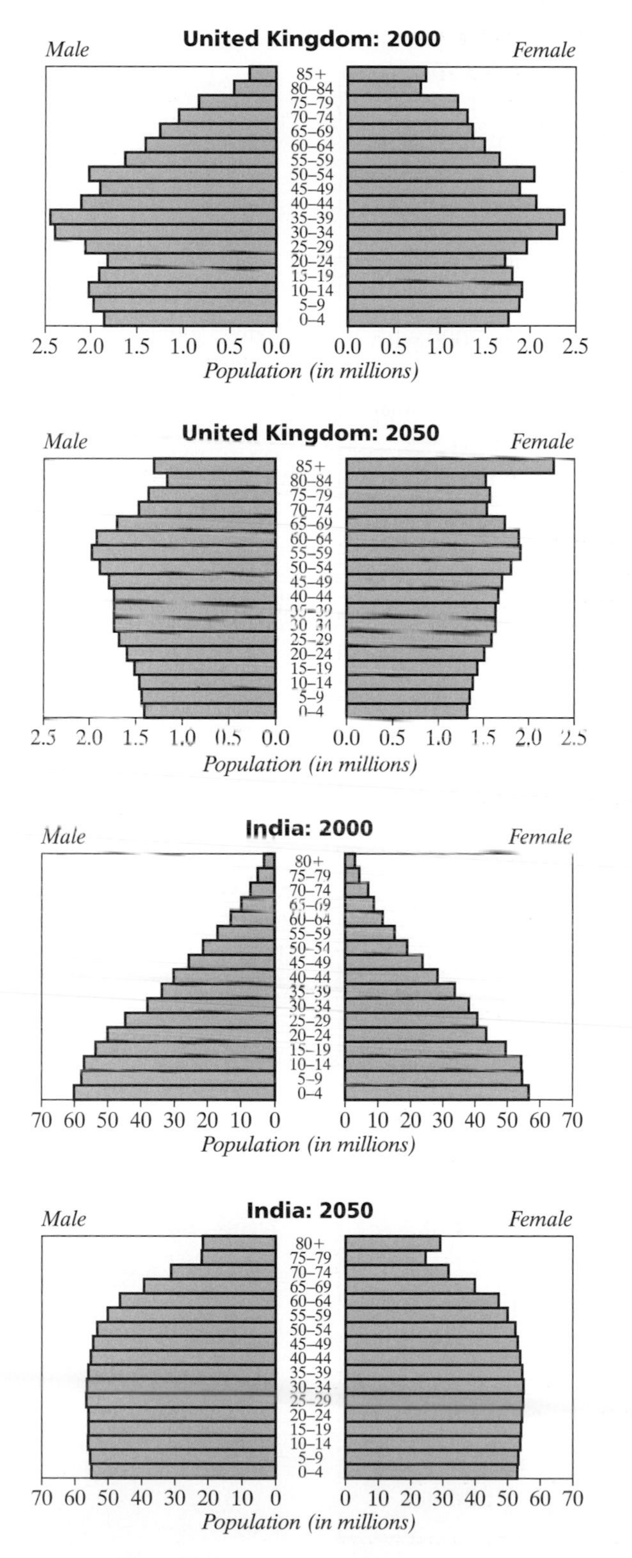

Discussion point

Look at the shapes of the pyramids, ignoring the numerical details. Describe the main features of each population distribution.

Activity 4.2B

1 Use the population pyramid for the UK in 2000.

a Estimate the total pre-school (0–4) population.

b Compare the male and female populations for people aged 80 years and above.

c i Identify the age group with the highest total frequency.

ii Calculate when the people in this age group were born.

iii Why might the birth rate have been high during these years?

iv What other reasons could contribute to the high number of people in this age group?

2 Use both UK population pyramids.

a i Compare the total pre-school population in 2000 with that predicted for 2050.

ii Compare the total population in the other age groups below 50.

iii What do your answers suggest will happen to the birth rate between 2000 and 2050?

iv What other reasons could there be for the differences between the population frequencies?

b i Estimate the percentage of the 30 to 34-year-old female population from 2000 expected to survive until 2050.

ii Estimate the percentage of the 30 to 34-year-old male population from 2000 expected to survive until 2050.

iii Suggest reasons for the difference between your answers to **parts bi** and **bii**.

c i Identify the modal age group for females in 2050.

ii Suggest reasons for the high frequency of this group.

3 Use the population pyramids for the UK and India in 2000.

a i What are the main differences between the population pyramids for the UK and India?

ii Suggest reasons for these differences.

b i Compare the number of boys between 0 and 4 years old with the number of girls between 0 and 4 years old in both countries.

ii What does this suggest about the ratio of male to female births in both countries?

4 Consider the projected population pyramid for India in 2050.

a Describe how it differs from the population pyramid for India in 2000 and give possible reasons.

b Compare the projected population pyramid for India in 2050 with that for the UK in 2050. Describe the similarities and differences. Comment on your findings.

Discussion point

Why is it only possible to give rough estimates when answering these questions, rather than more accurate values?

The **modal age group** is the age group with the highest frequency.

Note you are asked to
- describe the similarities and differences – to do this you should give factual information
- comment on your findings – now you should give some interpretation of the facts.

Candlestick charts

Candlestick charts are used by stock market traders to illustrate changes in market prices, identify patterns and use these patterns to predict what is likely to happen to investments in the short term future. They have been used in the Far East for over a century and since the early 1990s have also become a widely used method of market analysis in the UK.

A candlestick is drawn for each day by plotting the opening and closing price and the highest and lowest prices reached during the day. If the share price has increased over the day, a hollow (or white) rectangle is drawn between the opening and closing prices. If the share price has decreased a filled (black) rectangle is used instead. Lines are drawn above and below the rectangle to the points showing the highest and lowest share price during the day.

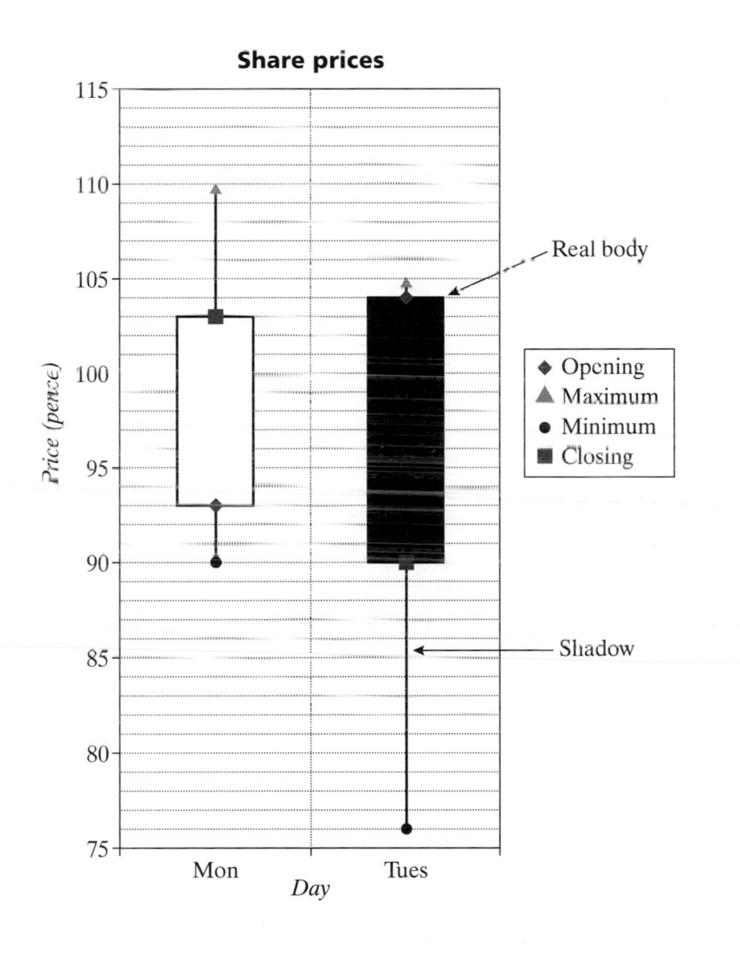

The *real body* is the rectangle drawn between the opening and closing prices of the day. The real body is *white* on days when the stock closes higher than it opens, and *black* on days when it closes lower than it opens.

The *shadows* are the vertical lines at each end of the real body. The *upper shadow* is the line drawn from the top of the real body to the day's highest price. The *lower shadow* is the line drawn from the bottom of the candlestick's real body to the day's lowest price.

In this example, Monday was a good day for owners of these shares. The price per share increased from 93p to £1.03.

During the day the price fell as low as 90p and reached as high as £1.10.

What happened on Tuesday?

Activity 4.2C

1 The candlestick chart for Wednesday, Thursday and Friday is shown below.

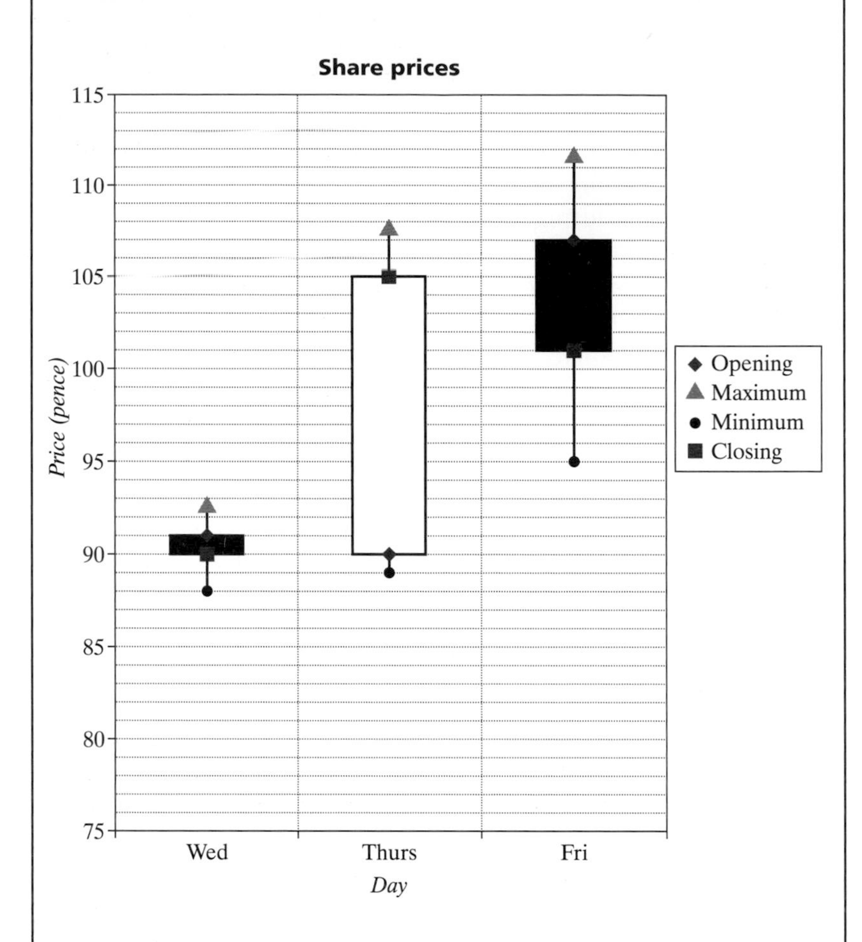

a Describe what happens on Wednesday, Thursday and Friday. For each day give the opening and closing prices as well as the highest and lowest prices reached during the day. Write two or three sentences for each day.

b Calculate the percentage change in the price for each day.

c Find the change in the price of the stock

- **i** between the stock market opening on Wednesday and closing on Friday
- **ii** between the beginning of the week (Monday) and the end of the week (Friday).

d Calculate the percentage change in the price

- **i** between the stock market opening on Wednesday and closing on Friday
- **ii** between the beginning of the week and the end of the week.

Discussion points

What does it mean if a candlestick

- has a long white real body
- has a short black real body
- has very long shadows
- has very short shadows?

What feature of the candlestick gives the range of prices during the day?

Discussion point

On which day do you think there was least trading in the shares?

2 Particular types of candlestick have special names. Look at the examples given below. In each case describe what happens to the price of the share during the day.

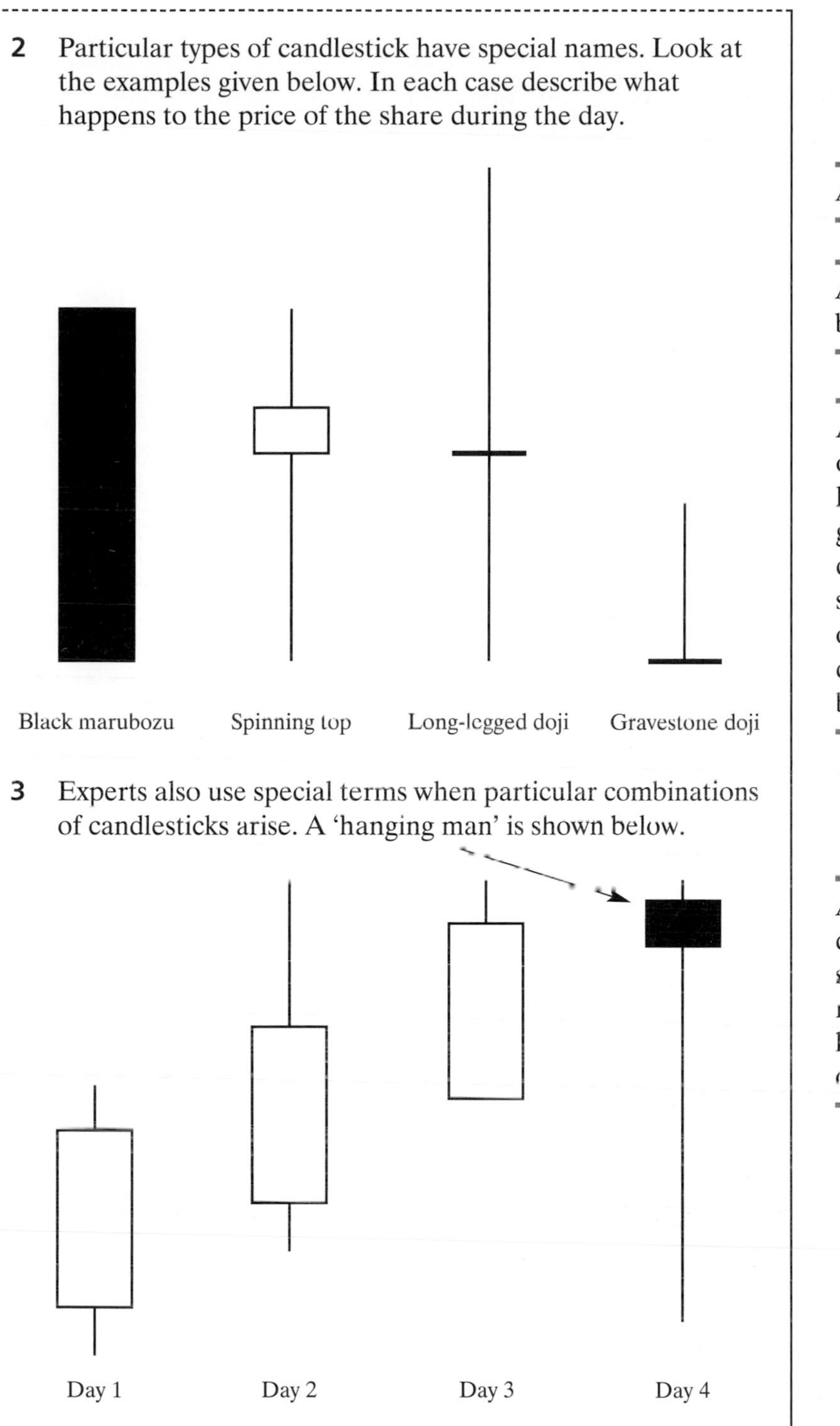

A *marubozu* has no shadows.

A *spinning top* has a small real body and long shadows.

A *doji* has a very small real body or no real body at all. The length of the shadows can vary, giving rise to different types of doji. A long-legged doji has long shadows – it is an extreme case of a spinning top. A gravestone doji has a long upper shadow but no lower shadow at all.

3 Experts also use special terms when particular combinations of candlesticks arise. A 'hanging man' is shown below.

A *hanging man* follows a period during which the price of the share has *risen*. It has a small real body (black or white), a long lower shadow and a short or non-existent upper shadow.

a Describe what happens to the price of the share when a black hanging man such as this occurs. What would you expect to happen next?

b What difference would it make to your answer to **part a** if the hanging man had been white rather than black?

c Sketch a sequence of candlesticks ending with a white hammer. Describe what has happened to the share price over the period you have shown.

A *hammer* looks exactly the same as a hanging man but is used after a period in which the price of the share has *fallen*.

4 Candlestick charts for Next and Marks and Spencer during June 1998 are given below.

Write an account of what happens to the share price of each company during this period. Describe any similarities and differences in their performances on the stock market.

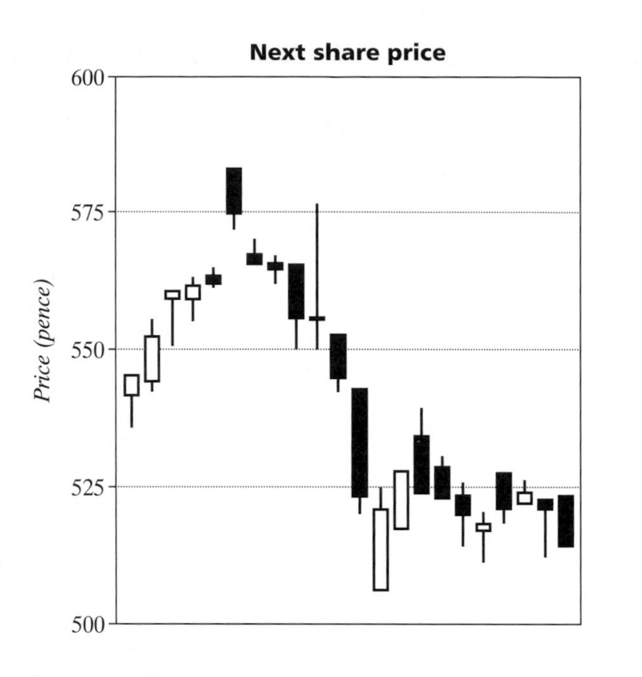

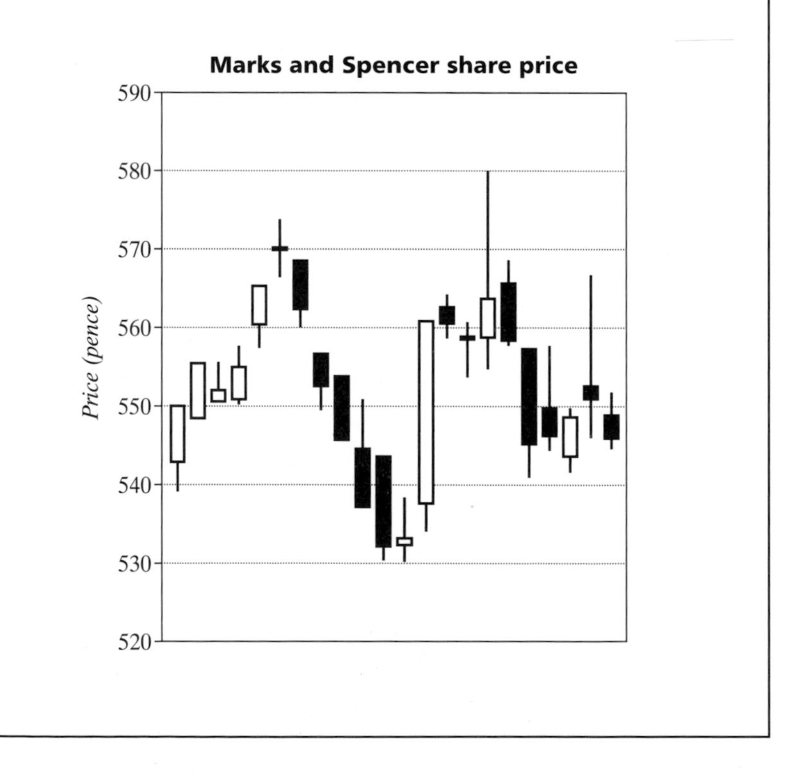

Interpreting statistical diagrams

Here you can explore more statistical diagrams that give a range of data in different contexts. These have been presented in the format of exam questions with data and question separated.

Aluminium cans

Data

The two bar charts give information about the consumption of aluminium cans in the UK and the number of cans recycled for the 10 years after 1988.

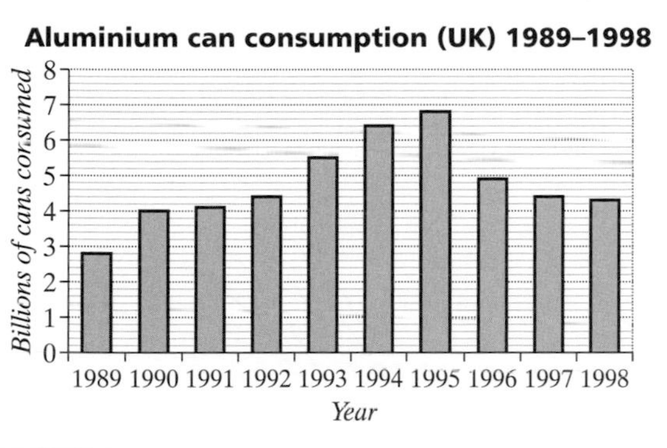

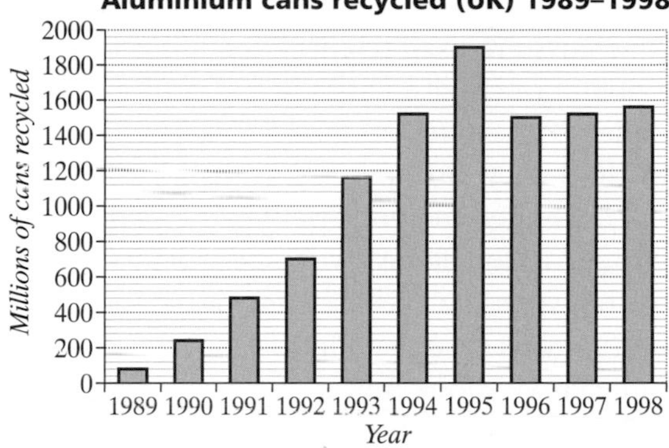

Question

1 Is the percentage of cans being recycled increasing or decreasing? Show how you made your decision.

2 Describe the general trends in the data.

Petroleum

Data

Figure 1 is a scatter diagram showing, for OECD[1] countries, consumption of petroleum plotted against petroleum-related emissions of carbon dioxide. Pearson's product moment correlation coefficient for the data is 0.999.

Figures 2 and **3** show, respectively, the petroleum-related carbon dioxide emissions for a sample of countries plotted against their Gross Domestic Product[2] and against their population.

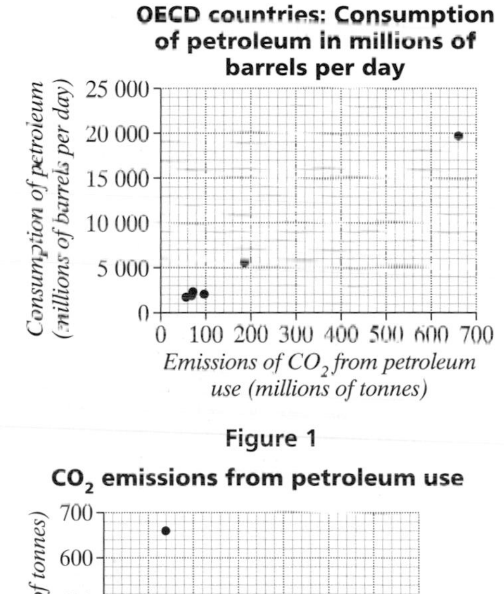

Figure 1

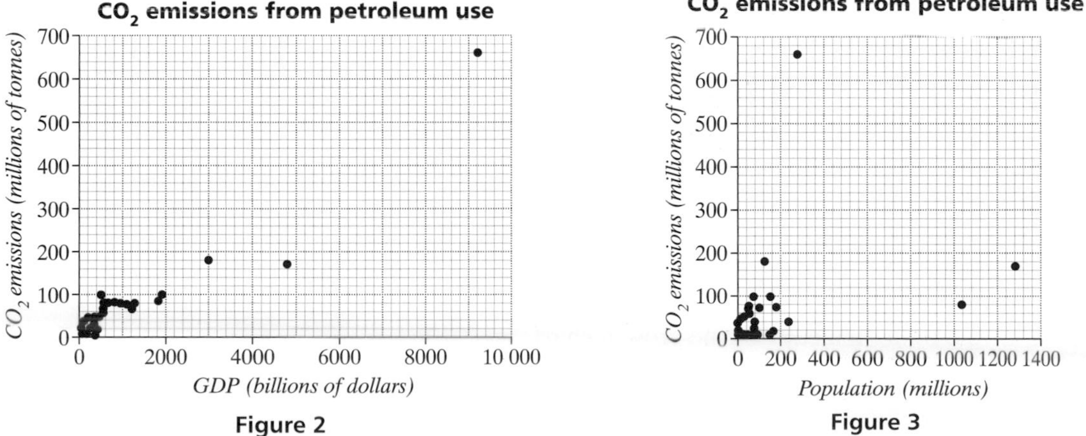

[1] OECD: Organization for Economic Cooperation and Development; an association of 21 nations to promote growth and trade.

[2] Gross Domestic Product (GDP); the total value of all goods and services produced domestically by a nation during a year.

Questions

3 What does the scatter diagram in **Figure 1** indicate about the connection between consumption of petroleum and petroleum-related carbon dioxide emissions for the OECD countries?

4 Describe the similarities and differences between the two scatter diagrams in **Figures 2** and **3**.

5 Write a paragraph or two in which you draw conclusions based on the three scatter diagrams.

Growth charts

Data

Growth charts are used to check the development of children as they grow older. The curves on this growth chart show how the 5th, 10th, 25th, 50th, 75th, 90th and 95th percentiles for height vary with age for boys between the ages of 2 and 20 years. In each case the 50th percentile curve shows how a 'typical' boy is expected to grow.

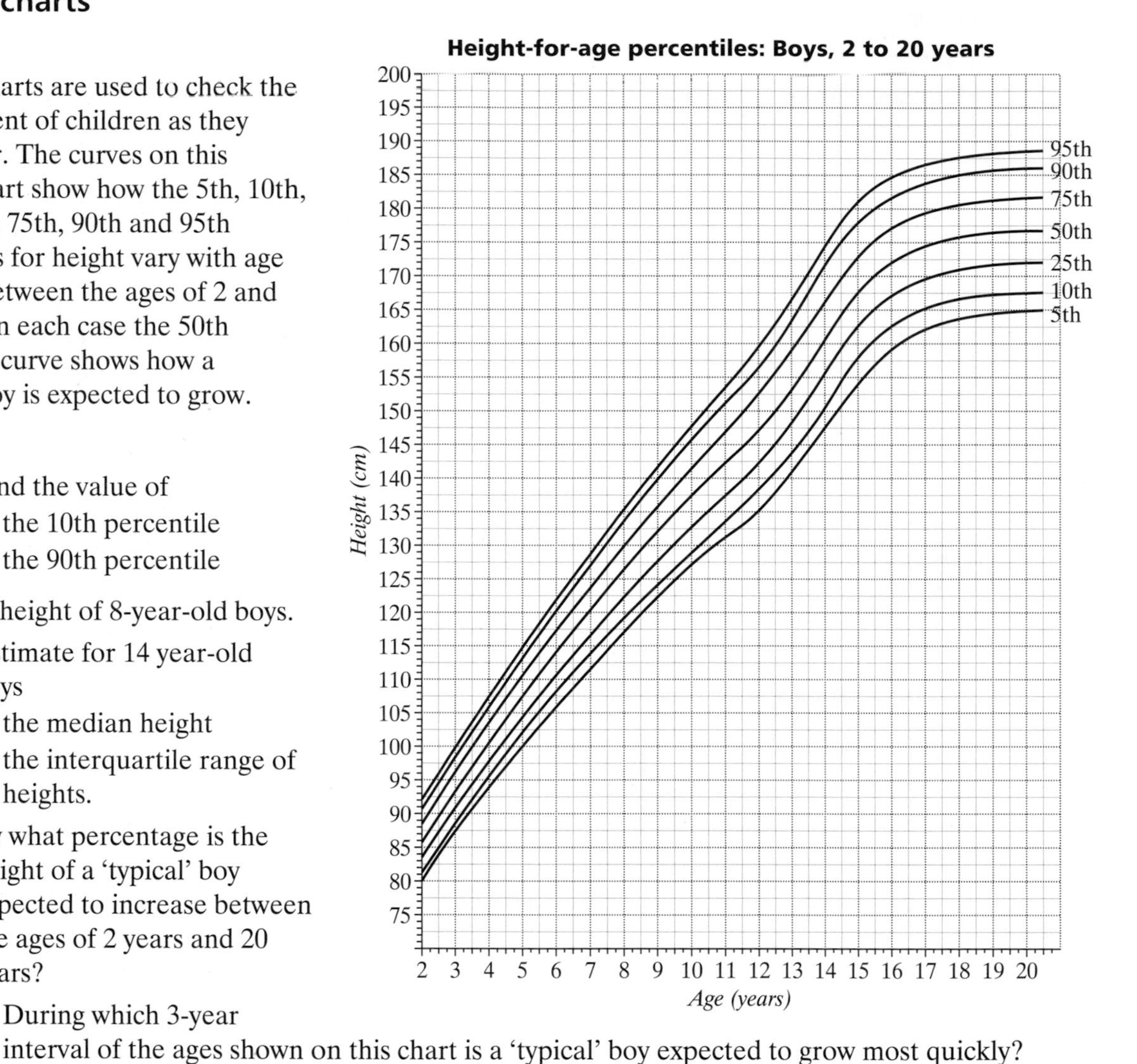

Questions

6 a Find the value of
 i the 10th percentile
 ii the 90th percentile

 of the height of 8-year-old boys.

 b Estimate for 14 year-old boys
 i the median height
 ii the interquartile range of heights.

 c By what percentage is the height of a 'typical' boy expected to increase between the ages of 2 years and 20 years?

 d i During which 3-year interval of the ages shown on this chart is a 'typical' boy expected to grow most quickly?
 ii What feature of the graph shows this?

7 The data below shows how Alastair's height increased from the age of 2 to 17.

Age	2	3	4	5	6	7	8	9	10	11	12	13	14	15	16	17
Height (mm)	847	954	1034	1092	1157	1223	1289	1351	1413	1463	1526	1576	1644	1717	1789	1819

 a Plot the data on a copy of the growth chart.

 b Write a paragraph explaining how Alastair's height varied in relation to that of the population remarking in particular on any variations from the 'typical'.

Stock market

Data

Charts like that shown below are used by stock market traders to analyse the performance of company shares. This chart shows how the price of British Airways shares varied in the six-month period from the beginning of May to the end of October 2002. As well as the closing price each day, the chart also shows a 50-day moving average of the share price and the number of shares sold each day.

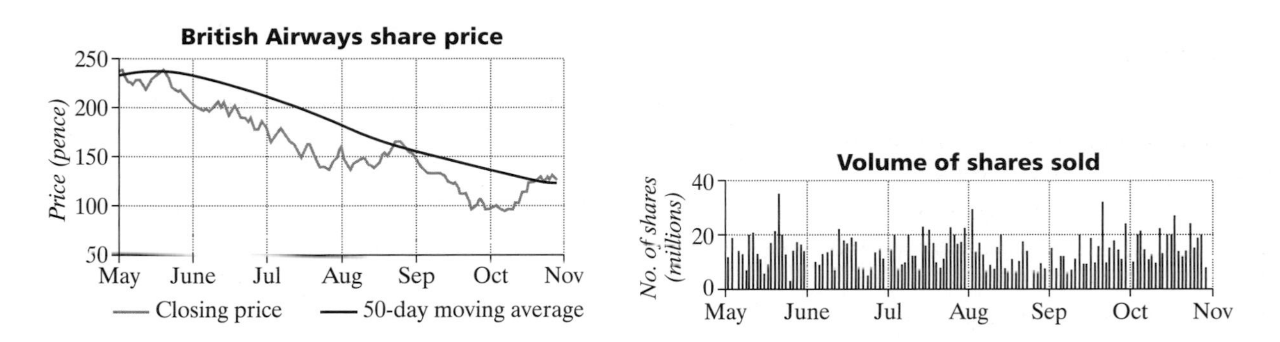

The **moving average** is a line on a stock chart indicating a long-term price trend with short-term fluctuations smoothed out. By comparing current prices with the smoothed long-term line, investors may read a buy or sell signal if current prices 'break through' the long-term trend line.

Question

8 a i Use the graph to estimate the minimum share price during this six-month period.
ii Give a reason why this might not be the minimum value at which the shares were sold during this period.

b Estimate the percentage fall in the price of shares between the beginning of June and the beginning of August.

c i Estimate the maximum volume of sales that occurred on one day in this period.
ii Estimate how much in pounds these shares cost.
iii Explain why there are fewer than 30 bars shown for each month in the volume chart.

d i Use the graph of the 50-day moving average to describe how the price of British Airways shares changed in the period shown.
ii Give one advantage of the graph of the 50-day moving average when compared with the graph of the daily closing prices.
iii Give one disadvantage of the graph of the 50-day moving average when compared with the graph of the daily closing prices.

The statistics of Florence Nightingale (1820–1910)

Data

Florence Nightingale used statistical data to draw polar area diagrams or 'coxcombs' as she called them. These showed deaths in British field hospitals in the Crimean War (1854–56).

In the diagram overleaf, area was proportional to the number of deaths. The central wedges showed deaths from wounds. The middle wedges represented death from 'other causes' and the outer wedges showed the deaths from contagious diseases – contracted by the soldiers while in hospital. The diagrams showed that the major cause of death was the unsanitary conditions of the hospitals!

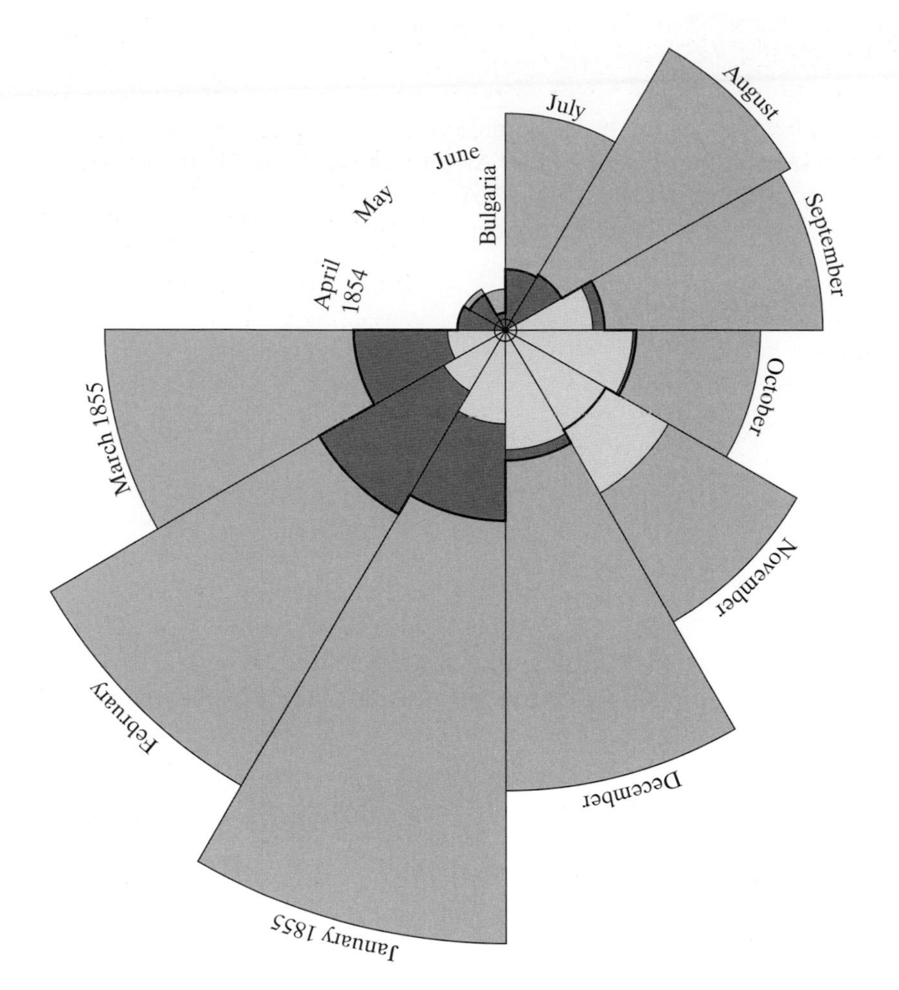

Florence Nightingale used this statistical analysis to campaign for the reform of sanitary conditions in hospitals.

In January 1855, 2761 soldiers died of contagious diseases, 83 from wounds and 324 from other causes.

Question

9 a The area of the total wedge representing January 1855 is $\frac{1}{12}$ of the area of the circle of radius 8 cm. (The area A cm^2 of a circle of radius r cm is given by $A = \pi r^2$). Show that the area of the wedge for January 1855 is 16.76 cm^2.

b Show that each square centimetre of Florence Nightingale's diagram represents the death of 189 soldiers.

c **i** Use your answer to **part b** to confirm that the area of the inner wedge for January 1855, representing 83 deaths from wounds should be 0.44 cm^2.

ii Calculate the radius of the wedge therefore required to represent deaths from wounds in January 1855. (Compare this calculated value with a value you measure on the diagram).

d Take measurements from the diagram to calculate

i total deaths in March 1855

ii deaths from contagious diseases in March 1885.

4.3 Revision summary

Introduction

Many different kinds of tables, charts and diagrams are used to present real-life data. Some of these are complex and require careful interpretation. A table or diagram may be used to show information about more than one event, such as a two-way table enabling comparisons between pass and fail rates for a test on two different occasions, or a graph showing the intended and the actual results of a production process so that quality can be controlled.

Statistical charts and graphs

Different groups of professionals, such as social scientists, manufacturers, stock market traders, etc. use the forms of presentation that suit their needs rather than those that may be familiar from most mathematics text books.

It's important to look carefully at what information is being presented: the use of two different scales on the vertical axis is common, as in the *subsidence* diagram, to allow comparison of two very different quantities year-by-year. In addition, the costs can be read off in either pounds or dollars.

Population pyramids are a common way of comparing the male and female populations of a country – and of comparing both of them with those of another country. These diagrams give, more effectively than a table of numbers, a quick and clear picture of the way in which the total population is split into the two genders and into different age bands.

Candlestick charts give several different pieces of financial information about daily share prices over a period of time, so traders experienced in using them can see at a glance what has happened to shares and, therefore, make predictions based on past performance.

Tables of figures are less easy to interpret, though experienced readers can spot trends by glancing across rows or up and down columns. Tables usually provide accurate numerical information, from which numbers can be extracted and used in calculations, such as percentage change calculations, to provide further information.

Such calculations can contribute to the general picture of a situation, such as the trends in drug trafficking in England and Wales. It's important to be clear what the percentages are percentages of.

The other tables presented offer you further opportunities to interpret complex information and suggest reasons for changes in data over time or between different areas.

Percentage increases and decreases

It is often easiest to use a multiplying factor to increase or decrease an amount by a percentage.

For example, to increase an amount by 4% calculate $1.04 \times amount$
to increase an amount by 0.2% calculate $1.002 \times amount$
to decrease an amount by 2.5% calculate $0.975 \times amount$.

If an amount is a previous amount increased by 3% then previous amount $= \frac{amount}{1.03}$

In an amount is a previous amount decreased by 17.5% then previous amount $= \frac{amount}{0.825}$.

4.4 Preparing for assessment

Your coursework portfolio

In this chapter you have looked at how you can work with, and interpret, information from tables of data and statistical diagrams. This work shows how you might analyse and interpret secondary data and also how you might present, analyse and interpret your own data using tables and diagrams.

An essential part of the process of carrying out a statistical enquiry is to interpret your findings making reference to the real situation. Although you need to be able to calculate quite complex statistical measures (for example, those you worked with in the last chapter) these are meaningless unless you can use them to inform you, and others, about what they tell you about the situation under investigation.

When writing up your reports for your coursework portfolio you must therefore make sure that you draw conclusions from every aspect of your presentation of statistical measures and diagrams. It is not enough to draw a histogram, for example. If you do this ask the questions:

- What is this telling me?
- What are the significant features?

If possible, contrast your findings with similar work carried out by others. For example, you may be able to find the results of a similar study using a search of the Internet.

In drawing conclusions you should highlight the limitations that you have identified in your work. For example, you may be concerned that your sample is not representative of the population you set out to consider, or that it is rather a small sample. It is important that you draw these limitations to the attention of the reader. If they have worked through the next chapter they will be casting a critical eye over your work. Perhaps you should leave finishing your own reports until you have worked through the next chapter yourself!

The Nuffield FSMQ website provides many data sets and web links that may be of use to you.

Practice exam questions

Hydrograph

Data

The hydrograph below shows rainfall during a rainstorm and discharge measured at a gauging station in a nearby river. Rainfall is usually measured in millimetres and discharge is measured in cubic metres per second (m^3s^{-1}).

Discharge that is not as a result of the storm is called baseflow. This can be identified in a variety of ways. This hydrograph uses the simplest method that assumes that baseflow is constant regardless of river depth.

The time between the highest rainfall recorded during the storm and the time it takes for the river to reach maximum discharge is known as the lag time.

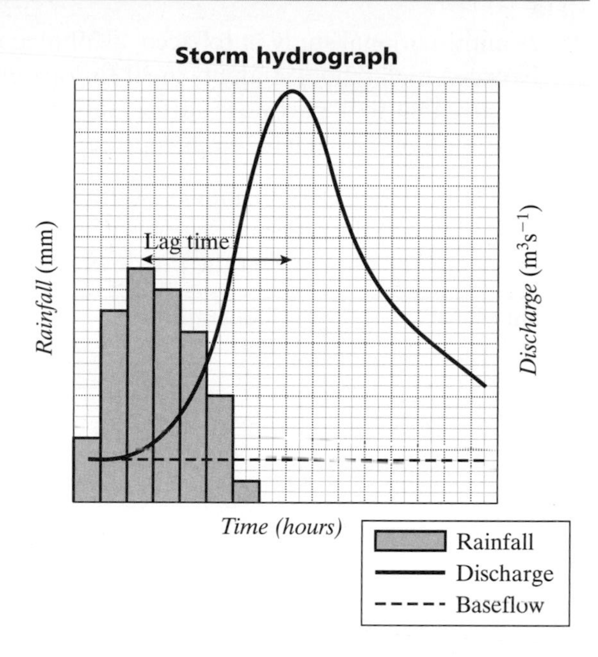

Question

1 The hydrograph is shown again below on a grid.

a Use the hydrograph to estimate

i the maximum discharge

ii the lag time

iii the time when the discharge was increasing most rapidly.

b i Calculate the total rainfall during the storm.

ii What was the mean rainfall per hour?

c Express the maximum discharge as a percentage of the baseflow.

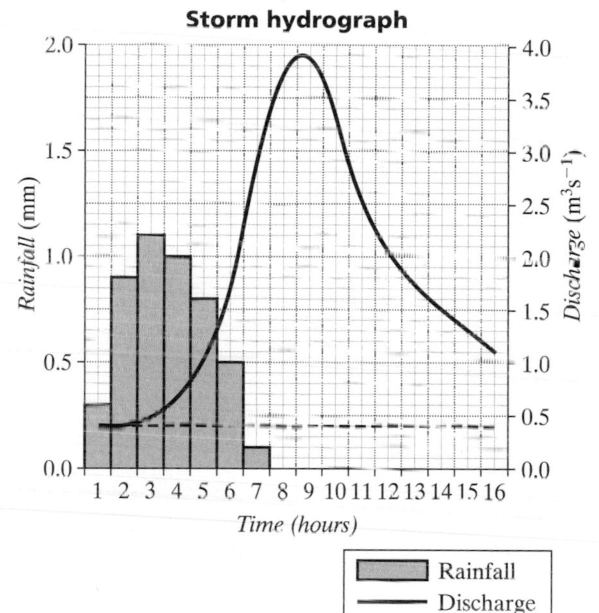

School survey – smoking and alcohol

Data

2 A multi-national study of tobacco, alcohol and marijuana use has found important differences between high school students in 30 European countries and in the United States. The survey in each country that took part was representative of the national student population in Year 10, with the exception of Russia which was representative of the capital of Moscow.

The table below gives estimates of the total Year 10 population in each country and the percentage of students in the national samples who said that they had smoked cigarettes, used marijuana or consumed alcohol during the previous 30 days.

Country	Y10 population (thousands)	Smoked cigarette(s) (%)	Used marijuana/ cannabis (%)	Drank alcohol (%)
Bulgaria	96.2	50	12	57
Croatia	56.3	38	16	46
Cyprus	9.5	16	27	61
Czech Republic	124.3	44	35	77
Denmark	64.4	38	24	85
Estonia	16.9	32	13	62
Faroe Islands	0.6	41	7	48
Finland	62.6	43	10	61
France	716.8	44	35	60
Greece	128.4	35	9	77
Greenland	0.7	67	23	59
Hungary	120.6	36	11	51
Iceland	3.4	28	15	43
Ireland	46.0	37	32	74
Italy	696.1	40	25	54
Latvia	29.3	40	17	58
Lithuania	44.7	40	12	73
Malta	4.7	32	7	75
Norway	54.1	40	12	55
Poland	467.1	33	14	61
Portugal	121.2	31	8	49
Romania	271.5	24	1	55
Russia	1760.4	45	22	63
Slovak Republic	65.3	37	19	60
Slovenia	24.1	29	25	62
Sweden	107.0	30	8	56
The Netherlands	192.0	36	28	66
Ukraine	599.8	40	20	56
United Kingdom	720.0	34	35	76
United States	3427.1	26	41	40
Yugoslavia	127.7	37	8	36

When grouped by region the results were:

Region	Smoked cigarette(s) (%)	Used marijuana/ cannabis (%)	Drank alcohol (%)
Northern Europe	39	19	62
Eastern Europe	38	16	58
Southern Europe	33	14	63
European average	37	17	61
United States	26	41	40

Question

a **i** The number of Year 10 students in the United States sample was 14 000.
What percentage is this of the total Year 10 population of the United States?

ii The survey in each country was said to be 'representative of the national student population in Year 10'. Explain briefly what this means.

b Estimate the total number of Year 10 students in the United Kingdom who

i smoked cigarettes

ii used marijuana

iii drank alcohol

in the 30 days preceding the survey.

c The percentages given in the table for Bulgaria add up to more than 100%. What does this tell you about the Bulgarian students who took part in the survey?

d **i** The second table shows that an average of 17% of Year 10 students in the participating European countries had used marijuana in the 30 days preceding the survey. Explain why this is not the same as the average of the percentages given for Northern, Eastern and Southern Europe

$$\left(\text{i.e. } \frac{19 + 16 + 14}{3}\right).$$

ii Use the results in the second table to describe two important differences between high school students in the European countries and in the United States.

5 Critical Thinking

▶ Contents

Do you agree with the students whose views are reported in this article from *The Independent*?

It is important that you think critically about the presentation of statistics.

For example, when reading this article you should be asking questions such as:

- Who was asked?
- Was the sample representative – did it include students from all types of schools, backgrounds and ages?
- What questions were asked – could these have affected the response of pupils?
- Do all students in all countries have the same understanding of the terms boring, time-wasting, bad behaviour, and so on?
- How are salaries measured?
- Is the proportion of graduates working in each country the same? If not, does this have an effect?

Bored, bored, bored: Why British teenagers feel lessons are all too often a waste of time

By Sarah Cassidy, Education Correspondent

30 October 2002

Britain's secondary school standards have fallen behind those in the rest of the developed world during the past 30 years, an international comparison published yesterday discloses.

The report also paints a disturbing picture of bad behaviour, time-wasting and boredom that it argues is now common in British classrooms.

Teenagers complain they have to put up with annoying levels of time-wasting, noise and disruption when they are trying to study.

British students also find school more boring than their peers in many of the 32 industrialised nations surveyed by the Organization for Economic Cooperation and Development (OECD). But if British students managed to progress to university they were likely to enjoy benefits graduates from other countries could only dream of, Andreas Schleicher, the OECD's head of statistics, said yesterday.

Degree holders in the UK saw the best return on their investment of any graduates in the countries analysed by his study, Mr Schleicher said. British graduates enjoyed significantly higher salaries than non-graduates and were much less likely to be unemployed, according to the OECD's 2002 edition of its annual *Education At a Glance* publication.

Perhaps the most important questions to ask are: Were these the most important findings reported in the publication? Has the newspaper selected the most significant findings?

Population who's going up, who's going down			
	1950	2000	2050
Australia	8.2	19.1	26.5
Bangladesh	41.8	137.4	265.4
Brazil	54.0	170.4	247.2
Canada	13.7	30.7	40.4
China	554.7	1275.1	1462.1
Denmark	4.3	5.3	5.1
France	41.8	59.2	61.8
Germany	68.4	82.0	70.8
India	357.6	1008.8	1572.1
Indonesia	79.5	212.1	311.3
Italy	47.1	57.5	43.0
Japan	83.6	127.1	109.2
Netherlands	10.1	15.9	15.8
Poland	24.8	38.6	33.4
Portugal	8.4	10.0	9.0
Russia	102.7	145.5	104.3
Spain	28.0	39.9	31.3
Sweden	7.0	8.8	7.8
UK	50.6	59.4	56.9
USA	157.8	283.2	397.1

After 200 years of continuous rapid population growth, there is little that inspires as much panic from political leaders, big business and right-wing populists as the prospect of population decline – which is imminent, according to the UN, in more than 60 countries.

Some countries, such as Japan, Russia and the Baltic states, have already fallen into the abyss. Italy's population and Germany's are shored up only by immigration. The recent British census showed population decline in Scotland and parts of Northern England. Across the UK as a whole, it could start as soon as 2020.

This extract from a *New Statesman* article by Anthony Browne accompanied the data alongside.

In this case questions that may come to mind might include:

- How are predictions of populations for the year 2050 made?
- What evidence is there of panic among political leaders?
- If there is population decline in Scotland and parts of Northern England, is this offset by population growth in other areas?

In this chapter, read three articles or reports which use statistics.

5.1 Reading between the lines

As you read each piece of statistics reported here, ask yourself the following questions.

- How was the sample chosen?
  Was the sample representative of the population?
  Was there any better way of selecting the sample?
- What questions were asked?
  Did they address the main points of the investigation?
  Were they clear, unambiguous, easy to understand and unbiased?
  Could the wording have been improved?
- What measurements were taken?
  Were they appropriate?
  How accurate are they?
- Were appropriate statistical charts and graphs used?
  Have they been drawn correctly?
  Are they misleading in any way?
  Would any other charts or graphs have been better?
- Were appropriate statistical measures calculated?
  Have they been calculated correctly?
  If the mean has been calculated, was it affected by outliers?
  Would any other statistical measures have been more appropriate?
- Do the conclusions reached correctly reflect the statistical evidence?

In many cases the report or article may not give sufficient information for you to be able to answer such questions. If this is the case, say so.

Youth crime

Read the article below, which was in *The Guardian* newspaper on 20 May 2002, following the release of a report compiled by MORI for the Youth Justice Board for England and Wales.

1 in 4 teenage pupils confess to crime

One in four teenage schoolchildren admit they have committed a crime in the past 12 months, according to a survey published last night by the government's Youth Justice Board.

The most common offence was fare dodging, and the detailed survey results suggested that levels of youth crime have remained relatively stable since 1999.

In the 2002 MORI youth survey, 26% of teenage pupils said they had committed a crime in the past 12 months. This compared with 24% in 1999.

The most common profile of a teenage offender was a white male aged 14–16, living in London or the North-East, who was excluded from school and had committed more than five crimes in the past year.

The survey revealed much higher levels of crime among teenagers who had been excluded from school, with 64% of this group admitting they had been involved in crime – down from 72% in 2000.

This group of teenagers was engaged in much more serious crime. Some 60% said they had been involved in handling stolen goods, 55% said they had carried a weapon other than a gun and 25% admitted they had stolen mobile phones.

The only crime in which there appeared to have been a drop was shoplifting, down from 58% to 49% among excluded children in the past year.

The results also showed that more schoolchildren who were caught were being punished. In 2001 some 22% said they were not punished after being caught for an offence. This figure fell to 16% last year.

But there were big variations around the country in the chances of a teenage offender being caught. Only 11% of those in London who admitted they had committed a crime said they had been caught, compared with 43% in Wales and 44% in the North-East.

Activity 5.1A

1 Is it clear what is meant by 'the past 12 months'?

2 What factors would you take into account when selecting a representative sample for the survey?

3 a How many teenagers would you include in the survey?
  b How many of these would be
    i schoolchildren
    ii excluded from school?
  c How would each sample be representative?

4 a What do you think would be the best way to carry out this survey? (For example, face-to face interview, telephone interview, questionnaire, ...).
  b Give reasons for your answer to **part a**.
  c Do you think the method would have any effect on the result?

5 What questions would you ask to find out this information?

6 Would some of the results have greater effect if they were shown graphically?
  a Is the report fair and unbiased?
  b Is the headline fair or sensationalist?

Below is some of the data from the MORI report on which the newspaper article is based.

Have you committed an offence in the last 12 months?

	School pupils				Excluded pupils		
	1999	2000	2001	**2002**	2000	2001	**2002**
Base:	3529	2767	5263	**5167**	*132*	*489*	***577***
YES	24%	22%	25%	**26%**	72%	60%	**64%**

What offence(s) have you committed in the last 12 months?

	Excluded pupils	
	2001	2002
Base:	*295*	*368*
Handling stolen goods	44%	60%
Carried a weapon (other than a gun)	50%	55%
Shop theft	58%	49%
Mobile phone theft	23%	25%

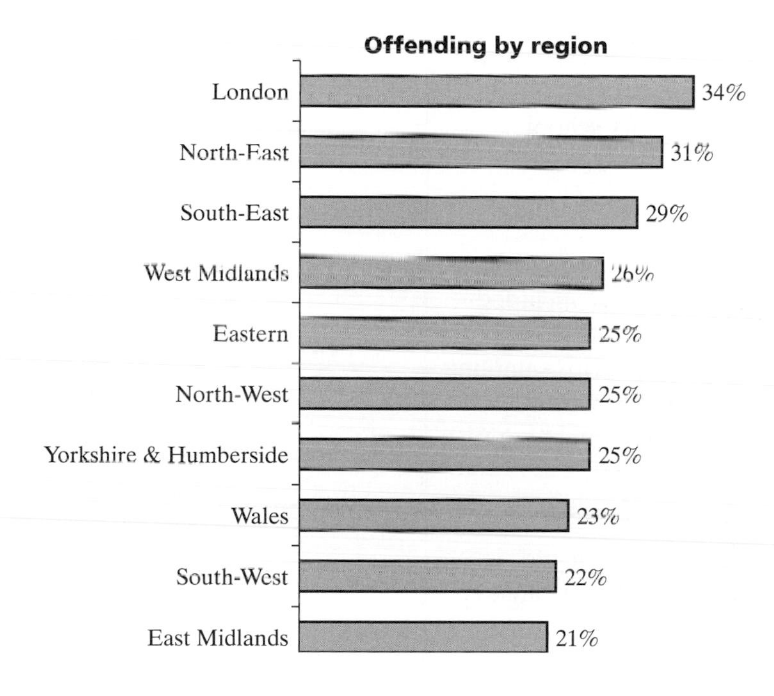

Activity 5.1B

1. Are the questions clear and unambiguous? Could they be interpreted differently by different pupils?
2. Are there any issues arising from the size of each sample?
3. How do the statisticians cope with the changing sample size from year to year?

4 The number of teenagers in the sample of those excluded from school in 2002 is more than one-tenth of the size of the sample of school pupils. Why might this be the case?

5 Why is the base of the 2002 excluded pupils who were asked about what offences they had committed 368 when the base of all excluded pupils is 577?

6 a Why might the rate of offending vary by region?
  b Why might rates of offending vary in a region from year to year?
  c What other information would you need to confirm or reject your hypotheses?

7 a What is the likely size of the sample in each region?
  b How would you select a sample of this size to be representative across a region?

Finally here, from the full MORI report for 2002, are some points that are made about how the survey was conducted.

Youth Survey 2002 for the Youth Justice Board
This report analyses the main findings from both the *2002 Survey of Secondary School Pupils*, and a parallel survey carried out among pupils excluded from mainstream secondary school education.

Survey of Secondary School Pupils
The sample of schools comprised 500 middle and secondary state schools in England and Wales. The sampling universe included county, voluntary aided/controlled and grant-maintained schools, but excluded special schools and 6th form colleges. This sampling frame was stratified by Government Office Region (GOR) and within each stratum, schools were selected proportional to the size of the school register, thus producing a nationally representative sample of secondary and middle schools.

The age groups included in the survey were 11–16-year-olds in curriculum years 7 to 11. Each school was randomly allocated one of these curriculum years, from which MORI interviewers selected one class at random (using a random number grid) to be interviewed. Interviewing was carried out through self-completion questionnaires with the whole class in one classroom period. A MORI interviewer was present to explain the survey to pupils, to reassure them about the confidentiality of the survey, to assist them in completing the questionnaire, and to collect completed questionnaires.

Fieldwork for the study was conducted between 14 January and 8 March 2002.

Data were weighted using a cell weight matrix of gender by age.

Survey of Excluded Pupils
Interviews were conducted among 577 children aged 11–16 in England and Wales currently attending a total of 82 project centres catering for excluded pupils.

Self-completion questionnaires were completed between 28 January and 12 March 2002. In certain cases, a trained interviewer assisted the respondent in completing the questionnaire by reading out the questions, and explaining difficult to understand words.

Data are un-weighted.

Comparison of Data Over Time
Where appropriate in the report, reference is made to results for the school surveys conducted on behalf of the Youth Justice Board in 1999, 2000 and 2001. Over time, some questions have been altered and therefore cannot be used to assess trends, particularly the list of offences used, which has been amended to ensure that the questions can be compared across the two 2002 surveys.

The trend data for the survey of *excluded children* should be treated with some caution. In 2000, excluded children who participated in the research were attending projects run by the voluntary organisation Include. This was expanded to encompass a range of funded education and employment projects, coordinated by a variety of organisations, in 2001. This year the sample for 2001 was used again, once again with additional projects.

The age ranges of excluded pupils have also been modified over time to enable comparisons to be made between the MORI *Schools Omnibus Survey* and the survey of excluded children. In 2000, excluded pupils aged 14–18 were surveyed, but this was changed to 11–16 year-olds in 2001, and has remained thus in 2002. Therefore, comparisons can be made between 2001 and 2002, but it should be borne in mind that the figures for 2000 relate to older pupils only.

It should be noted that the profile of children between the two surveys is very different. For example, excluded pupils tend to be boys, and aged 15–16.

When interpreting the findings it is important to remember that the results are based on a sample, not the entire population. Consequently, results are subject to sampling tolerances, and not all differences between sub-groups are statistically significant.

Activity 5.1C

1 Should special schools and 6th form colleges be excluded if you want a representative sample?

2 Think about the way each class was chosen for interview. What effect might this have on the representativeness of a sample for any of the regions?

3 Secondary school pupils were surveyed between 14 January and 8 March. What implications has this in terms of the question 'Have you committed an offence in the last 12 months?'

4 What is a cell weight matrix of gender by age?

5 All excluded pupils were attending project centres. Does this allow the samples to be representative?

6 Should the data for excluded pupils be weighted?

7 Over time, the wording of some questions has been altered. What effect might this have on results?

8 What effect might the way data altering is processed have on results (for example, considering different lists of offences)?

9 Excluded pupils from additional organisations have been included over time. Should the changing base of a sample be accounted for by ensuring that the sample is always representative?

10 a The age range of excluded pupils surveyed has changed over time. What effect might this have?

b Could the original data be reanalysed so comparisons can be made?

11 Do any of the points from the MORI report significantly affect the newspaper article?

Build our bypass

A bypass has been proposed to alleviate traffic congestion in the villages of Ridgmont and Husbourne Crawley. A fictitious leaflet produced by an imaginary action group to support the bypass is given on the following pages.

BUILD OUR BYPASS

Do you want:

- good health and clean, fresh air to breathe?
- a quieter life, free from the noise of road traffic?
- your children to be safe crossing the road?
- to rid our village of juggernauts?

If so join BOB (Build Our Bypass) in the fight for our bypass. Protesters are trying to have plans for the bypass dropped. *If you do not take action now to support the bypass plan, you will regret it later.*

Plans for the bypass

As you will know only too well, *traffic in Ridgmont is getting worse by the day*.

The proposed Ridgmont Bypass is shown on the map below. It is in two sections.

The East/West section will remove 70% of the traffic that presently passes through Ridgmont on its way to Junction 13 on the M1. The North/South section will remove another 20% of Ridgmont traffic as well as 60% of traffic in Husborne Crawley.

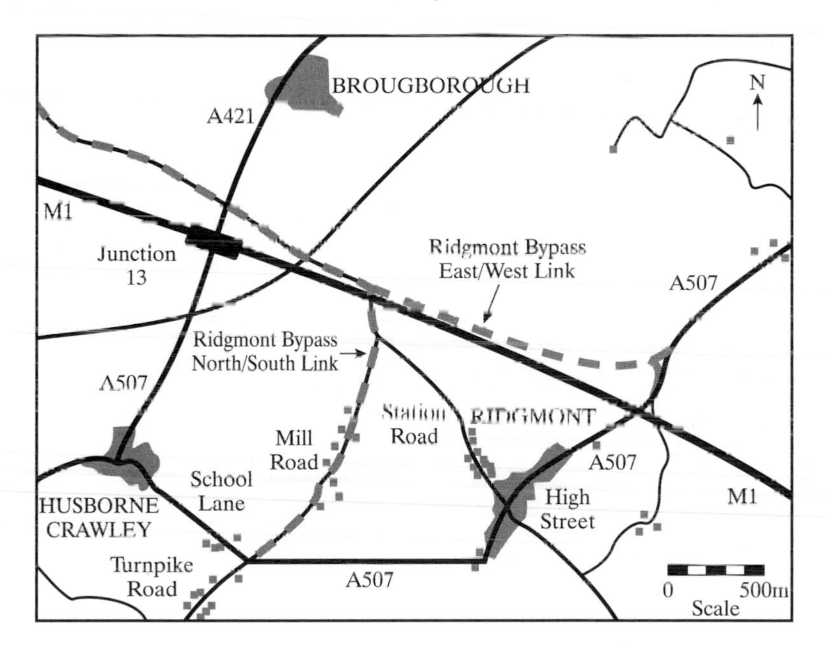

What has BOB done so far?

As soon as plans for the bypass were published it became clear that there would be opposition from environmental groups and others. *The huge majority of opposers do not live in this area* and have no idea about how traffic through Ridgmont blights our lives. *BOB is an action group set up by people living in the Ridgmont area to fight for our bypass*. Over the past year BOB has been collecting evidence of the need for our bypass. We have carried out surveys and gathered a wide variety of information to present at a public inquiry. This leaflet gives details of our work.

What villagers think

BOB has consulted everyone who lives in Ridgmont. Five public meetings have been held and over 1100 people returned a questionnaire about the bypass plans. We asked what people thought of the bypass plans, how the bypass would improve their lives if it was built and what worries they would have if it did not go ahead. *More than 90% of Ridgmont residents broadly agree with the draft plan and think that a bypass will improve their quality of life significantly.*

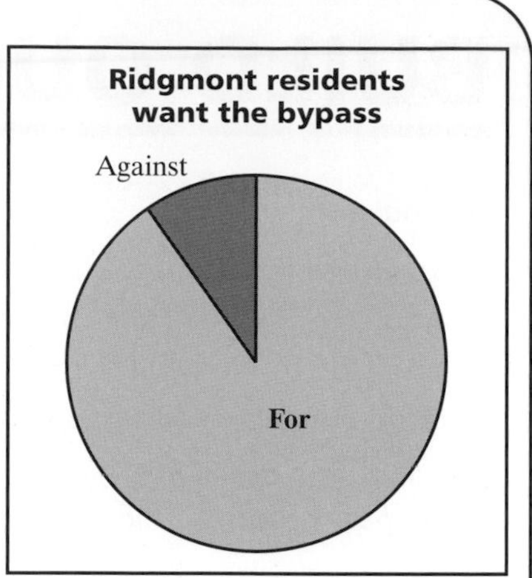

If the bypass is not built, most people are worried about the effect road traffic will have on the health and safety of the people who live here. Some people also fear that the value of their houses will fall.

As well as the questionnaires, BOB has also received over 1200 letters and petitions with 3900 signatures, mostly in favour of the bypass.

BOB's researchers have been at work finding out the facts about the issues raised. Evidence of the problems caused by traffic congestion is given later in this leaflet.

What drivers think

From 8 am to 9.30 am and 4.30 pm to 6.30 pm each day for a week, BOB's volunteers counted vehicles and distributed questionnaires to all traffic passing through Ridgmont. The drivers were asked where they had started their journey, their destination and whether or not they had or would be using Junction 13 on the M1. We found that only 12% of traffic was local and that 72% of drivers were passing through Ridgmont before or after using the motorway.

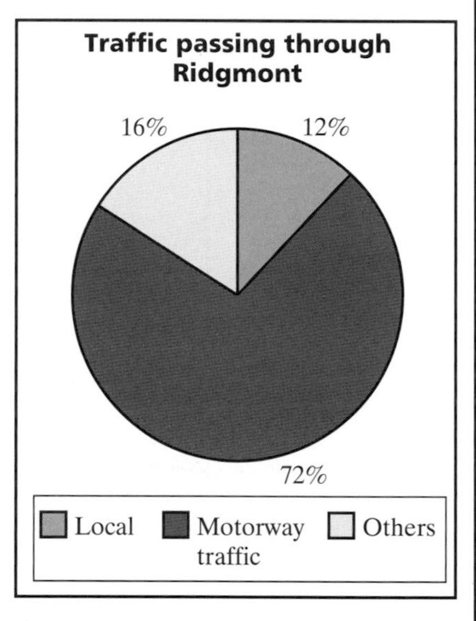

Drivers were also asked to rate, on a scale of 1 to 10, the stress they experienced as a result of the congestion in Ridgmont, whether or not they would be in favour of a bypass and how much time they estimated it would save them. The average stress level was rated at 9, giving a clear indication that *a drive through Ridgmont is a stressful experience*. It is a well-known fact that stress causes poor driving, makes a driver less observant and in extreme cases can cause accidents or result in road rage.

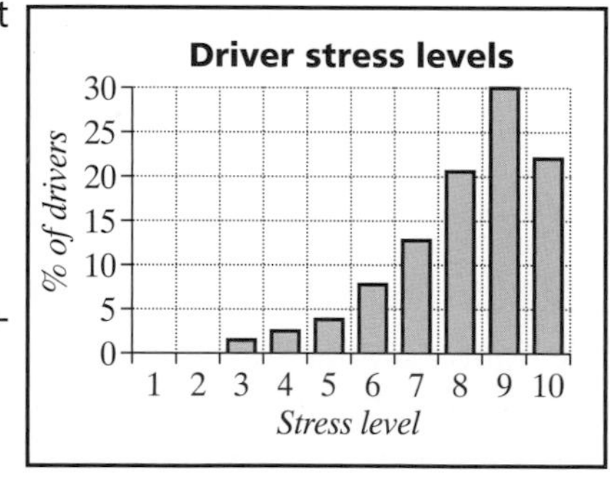

Not surprisingly, *every driver who returned the questionnaire was in favour of a bypass*. The time they expected to save was on average 10 minutes. This may not seem a lot, but as the average number of vehicles per hour passing through Ridgmont is over 1000 this gives an estimated total time of over 4000 man-hours per week lost because of congestion in Ridgmont. This is over 2 years of work.

The 10 minute delay caused at present by traffic congestion has an effect on the local bus service causing people to be late for work or appointments when they travel at peak times. A bypass would shorten your journey time significantly. If you want to make travelling in this area quicker and more enjoyable, **fight for our bypass!**

Noise and pollution

Everyone who lives in Ridgmont is fed up with the noise, pollution and congestion caused by too many HGVs and cars passing through our village.

BOB has taken sound and pollution measurements on the A507 in the centre of Ridgmont and also in Beachampton, a village of a similar size just a few miles away. On a typical workday the noise level regularly reaches 75 decibels in Ridgmont whereas in Beachampton it rarely gets as high as 70 decibels.

HGVs on our roads.

These figures may not seem very different, but because of the way sound is measured they show that traffic noise in Ridgmont is a staggering *4 times as bad* as the noise in Beachampton.

Traffic is also responsible for high levels of nitrogen dioxide, carbon monoxide and other air pollutants. Such *pollution causes respiratory diseases such as asthma and bronchitis*. In fact scientists in Nottingham have recently studied how the incidence of asthma increases as you get nearer to a major road. For primary school children the incidence increases by 6% for every 30 metres, and for secondary school children the incidence increases by 16% for every 30 metres.

BOB's researchers have taken measurements of nitrogen dioxide and carbon monoxide levels during peak periods. The nitrogen dioxide results for the last six months are shown on the graph.

The pollution in our air is steadily getting worse.

Nitrogen dioxide levels will soon regularly reach 200 micrograms per cubic metre.

The government's target is for the nitrogen dioxide level not to exceed 200 micrograms per cubic metre more than 18 times per year.

It will not be long before Ridgmont fails to meet the target.

The noise and pollution caused by traffic is making life a misery for people living near the A507 in Ridgmont. If you value your children's health, **fight for our bypass!**

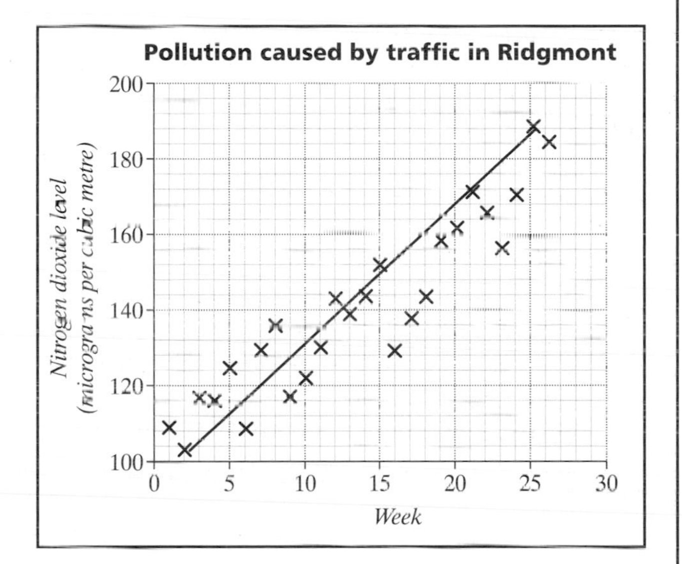

House prices

The graph shows how the average price of houses in Ridgmont compares with prices in Beachampton, just a few miles away.

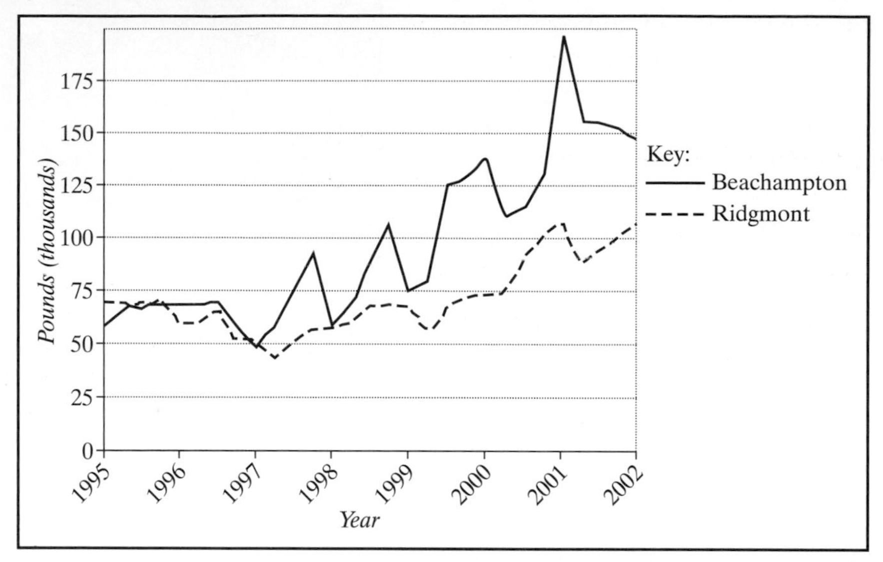

As you can see, prices were similar from 1995 to 1997, but recently Ridgmont has fallen further and further behind. The table shows that in *Ridgmont your house is losing value no matter what size of house you own*.

	Beachampton		Ridgmont	
Detached	£188 072	(38)	£152 203	(11)
Semi-detached	£109 908	(30)	£97 285	(7)
Terraced	£92 315	(10)	£73 969	(13)
All property types	£145 732	(78)	£106 994	(31)

Brackets give the number of sales.

It does not take a genius to realise that the reason for the price difference between these villages is the fact that Ridgmont has a busy road running through the middle of it whereas Beachampton doesn't. If you want to increase the value of your house, **fight for our bypass!**

What can you do to help fight for our bypass?

It is vital that widespread public support for the bypass is demonstrated clearly to all concerned. To help you can:

- study this leaflet to ensure that you are well-informed about the main arguments
- visit the exhibition at the County Council Offices where Officers will be available to answer questions and listen to your views on the scheme
- write to the Highways Agency and the Department of Transport who will make the final decision on whether or not the scheme will go ahead
- display BOB posters in your house and car
- join our rallies and events (details will be published in the local press and announced on local radio and television).

Activity 5.1D

Write a report in which you critically examine the information given in the leaflet. Make sure you consider and answer each of the questions at the start of this chapter (page 120 and 121).

Difference of opinion

Sometimes different groups of people can come to completely different conclusions based on the same set of statistical results.

For example, in 1998 the World Health Organization (WHO) completed a research project on passive smoking in Europe. Below is a report from the *Sunday Telegraph* which was printed on 8 March 1998 before the WHO published their report.

Passive smoking is the inhalation of tobacco smoke from other people's cigarettes.

Passive smoking doesn't cause cancer – official

by Victoria Macdonald, Health Correspondent

The world's leading health organisation has withheld from publication a study which shows that not only might there be no link between passive smoking and lung cancer but that it could even have a protective effect.

The astounding results are set to throw wide open the debate on passive smoking health risks. The World Health Organization, which commissioned the 12-centre, seven country European study has failed to make the findings public, and has instead produced only a summary of the results in an internal report.

Despite repeated approaches, nobody at the WHO headquarters in Geneva would comment on the findings last week. At its International Agency for Research on Cancer in Lyon, France, which coordinated the study, a spokesman would say only that the full report had been submitted to a science journal and no publication date had been set.

The findings are certain to be an embarrassment to the WHO, which has spent years and vast sums on anti-smoking and anti-tobacco campaigns. The study is one of the largest ever to look at the link between passive smoking – or environmental tobacco smoke (ETS) – and lung cancer, and had been eagerly awaited by medical experts and campaigning groups.

Yet the scientists have found that there was no statistical evidence that passive smoking caused lung cancer. The research compared 650 lung cancer patients with 1,542 healthy people. It looked at people who were married to smokers, worked with smokers, both worked and were married to smokers, and those who grew up with smokers.

The results are consistent with there being no additional risk for a person living or working with a smoker and could be consistent with passive smoke having a protective effect against lung cancer. The summary, seen by *The Telegraph*, also states: 'There was no association between lung cancer risk and ETS exposure during childhood.'

A spokesman for Action on Smoking and Health said the findings 'seem rather surprising given the evidence from other major reviews on the subject which have shown a clear association between passive smoking and a number of diseases'. Roy Castle, the jazz musician and television presenter who died from lung cancer in 1994, claimed that he contracted the disease from years of inhaling smoke while performing in pubs and clubs.

A report published in the British Medical Journal last October was hailed by the anti-tobacco lobby as definitive proof when it claimed that non-smokers living with smokers had a 25 per cent risk of developing lung cancer. But yesterday, Dr Chris Proctor, head of science for BAT Industries, the tobacco group, said the findings had to be taken seriously. 'If this study cannot find any statistically valid risk you have to ask if there can be any risk at all.

'It confirms what we and many other scientists have long believed, that while smoking in public may be annoying to some non-smokers, the science does not show that being around a smoker is a lung-cancer risk.'

Action for Smoking and Health (ASH) is a British organisation that aims to *'preserve the health of the community; to advance the education of the public concerning smoking; to carry out or support research and communication for the benefit of the health of the community'.*

BAT stands for British American Tobacco.

WHO were quick to reply. The press release they issued on 11 March 1998 is given below:

PASSIVE SMOKING DOES CAUSE LUNG CANCER, DO NOT LET THEM FOOL YOU

The World Health Organization (WHO) has been publicly accused of suppressing information. Its opponents say that WHO has withheld from publication its own report that was aimed at but supposedly failed to scientifically prove that there is an association between passive smoking, or environmental tobacco smoke (ETS), and a number of diseases, lung cancer in particular. *Both statements are untrue.*

The study in question is a case-control study on the effects of ETS on lung cancer risk in European populations, which has been carried out over the last seven years by 12 research centres in 7 European countries under the leadership of WHO's cancer research branch – the International Agency for Research on Cancer (IARC).

The results of this study, which have been completely misrepresented in recent news reports, are very much in line with the results of similar studies both in Europe and elsewhere: *passive smoking causes lung cancer in non-smokers.*

The study found that there was an estimated 16% increased risk of lung cancer among non-smoking spouses of smokers. For workplace exposure the estimated increase in risk was 17%. However, due to small sample size, neither increased risk was statistically significant. Although, the study points towards a decreasing risk after cessation of exposure.

In February 1998, according to usual scientific practice, a paper reporting the main study results was sent to a reputable scientific journal for consideration and peer review. That is why the full report is not yet publicly available. Under the circumstances, however, the authors of the study have agreed to make an abstract of the report available to the media.

'It is extremely important to note that the results of this study are consistent with the results of major scientific reviews of this question published during 1997 by the government of Australia, the US Environmental Protection Agency and the State of California,' said Neil Collishaw, Acting Chief of WHO's Tobacco or Health Unit in Geneva. 'A major meta-analysis of passive smoking and lung cancer was also published in the *British Medical Journal* in 1997. From these and other previous reviews of the scientific evidence emerges a clear global scientific consensus – passive smoking does cause lung cancer and other diseases,' he concluded.

'IARC is proud of the careful scientific work done by the European scientific team responsible for this study,' commented Dr Paul Kleihues, the Agency's director. 'We are very concerned about the false and misleading statements recently published in the mass media. It is no coincidence that this misinformation originally appeared in the British press just before the No-Tobacco Day in the United Kingdom and the scheduled publication of the report of the British Scientific Committee on Tobacco and Health.'

In a briefing ASH explained how the *Sunday Telegraph* were distorting the facts:

How the *Sunday Telegraph* and BAT got it badly wrong on passive smoking

The (WHO) study gives the following figures for relative risk of a non-smoker contracting lung cancer as a result of living with a smoking spouse or working in a smoking workplace.

Spouse smokes 1.16 Smoky work place 1.17

These figures mean that non-smokers are 16% more likely to get lung cancer if their spouse smokes than if they live with a non-smoking spouse. In practice this is a small risk – an active smoker has a risk 20 times a non-smoker and a non-smoker's risk of getting lung cancer is small in absolute terms. However, applied to the millions of people in this situation, the 16–17% extra risk would amount to an extra several hundred deaths per year in the UK. The WHO has subsequently rounded on BAT and the newspaper, saying its reporting of the findings was 'false and misleading' and 'From these and other previous reviews of the scientific evidence emerges a clear global scientific consensus – passive smoking does cause lung cancer and other diseases' (press release – 9 March). Despite this, BAT and tobacco industry spokespeople continue to push the *Sunday Telegraph*'s misleading version.

So how did BAT/*Sunday Telegraph* manage to make this into a '*Passive smoking doesn't cause cancer – it's official*' headline?

The error (or deception) was to misinterpret a statistical test applied to these results. Because the estimate of risk is based on a sample of 650 lung cancer cases, the risk in the whole population might be different because the sample may not be exactly representative. So the statisticians develop a 'confidence interval'.

This allows them to give the central estimates above and then say 'we are 95% confident that the real value for the whole population lies between

British American Tobacco and other tobacco companies were suspected of taking part in a coordinated attempt to discredit the WHO research.

the following limits x and y' – i.e. the chance that the actual risk lies outside this range is 1 in 20 or 5%. For the WHO study the limits are as follows:

	Central Estimate	Lower Limit (x)	Upper Limit (y)	Number of Cases
Spouse	1.16	0.93	1.44	650
Workplace	1.17	0.94	1.45	
BMJ	1.24	1.13	1.36	4626

The fact that the lower limits drop below 1.0 shows that the statisticians cannot be 95% confident that the survey has detected an effect – it is possible to obtain the central estimate of 1.16 and 1.17 by chance but for there to be no real effect. This is what the statisticians mean when they say the result is not 'statistically significant' – i.e. they cannot be 95% certain that they have detected a link between passive smoking and lung cancer. However, this uncertainty was inverted and reported as evidence of 'no effect'. The tobacco industry has translated this formal statistical meaning of the word 'significance' into lay language and has been arguing that 'the study shows the risk is insignificant' (paraphrase).

Furthermore, because the lower limit is 0.93 it was translated to a possible 'protective effect'. Of course the study no more shows a protective effect than it shows a 44% increase in risk – the other extreme of the confidence interval. This is an outrageous misinterpretation of the results and it is difficult to know if this was naivety on the part of the *Sunday Telegraph* or manipulation by BAT, who should know better, or both.

The figures from the BMJ report of 18th October 1997 (Vol 315 p 980–988) are added above for comparison. This shows that the ranges overlap and therefore that the results are consistent. The BMJ results have a much smaller confidence interval – i.e. the use of several studies in a 'meta-analysis' increased the effective sample size, and therefore the overall sample is more likely to be representative. Note, it may seem paradoxical at first sight, but when several studies that each show no statistical significance are added together, the overall result may be statistically significant – each of the studies contributes to greater confidence in detection of the effect and a larger overall sample and hence smaller confidence interval. This is why meta analysis is important and used to detect what are quite small additional risks.

In no way can the illustration of the WHO results above be used to support the thesis that there is no effect or that there is a protective effect. The result does point towards a link between passive smoking and lung cancer because it is consistent with other major studies. There are also sources of evidence other than epidemiology that support the argument.

Finally, it should be noted that the choice of a 95% confidence interval is arbitrary, though conventionally used as a scientific test. However, when policy makers have to make decisions, they cannot reserve judgement because doing nothing is to act as if there is no risk – and there is no evidence to support that idea. For this reason, a policy maker without the luxury of reserving judgement would be right to use a lower test of statistical significance ... we would be unsettled if the Government failed to act on passive smoking if it was 80% sure that it causes lung cancer – even if wasn't 95% sure. The 80% confidence interval would be smaller and probably show a statistically significant link between passive smoking and lung cancer in the WHO study.

To be precise, when a 'statistical significance' is quoted it should be accompanied by the confidence interval – i.e. 'the results were not statistically significant at the 95% level'. But then we doubt that being precise was what BAT had in mind.

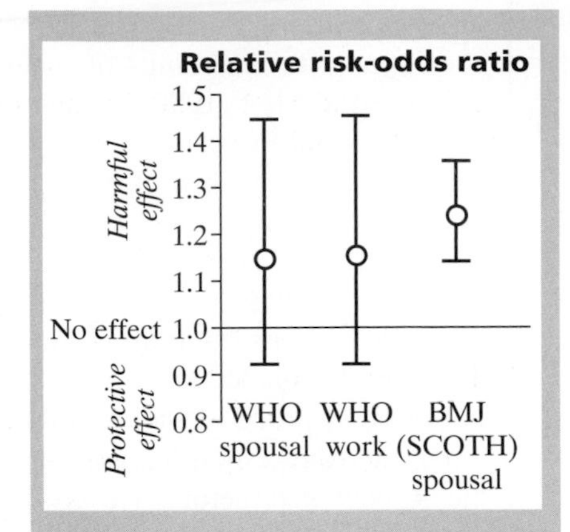

This diagram was included in the ASH briefing to illustrate these results.

The bar showing the results reported by the *British Medical Journal* (BMJ) indicates that they are 95% confident that the relative risk of contracting lung cancer from passive smoking is between 1.13 and 1.36, (corresponding to an increase of between 13% and 36%).
The central point at 1.24 indicates that their best estimate of the increase in cases of lung cancer due to passive smoking is 24%.

Activity 5.1E

1. What are the main points made by the *Sunday Telegraph* on 8 March?
2. a What are the main points of response made in the WHO's press release of 11 March?
   b What information is given here that was not given in the *Sunday Telegraph*'s report?
3. a What do WHO mean by *'However, due to small sample size, neither increased risk was statistically significant'* mean?
   b How was this interpreted by the *Sunday Telegraph*?
4. Compare the reason for the delay in publication of the research results given in WHO's press release with the reason suggested in the original newspaper article.
5. a According to ASH's briefing, more weight should be given to the BMJ results than to the WHO results. Why is this?
   b Look at ASH's 'Relative risk – odds ratio' chart. Why is the BMJ bar shorter than the others?
6. a Explain in your own words how the *Sunday Telegraph* would argue that the WHO results not only show *'that there might be no link between passive smoking and lung cancer but that it could even have a protective effect'*.
   b Is this interpretation of the results reasonable?

Activity 5.1F (optional)

By 1998 it had become widely accepted that smoking itself contributed to death rates from lung cancer and other diseases, but arguments raged over just how many deaths it was responsible for.

In its *Morbidity and Mortality Weekly Report* on 27 August 1993 the US Center for Disease Control (CDC) suggested that each year over 400 000 Americans die from diseases caused by smoking.

In 1998 Robert Levy and Rosalind Marimont argued that this estimate of 400 000 deaths was grossly exaggerated and scientifically unsound (in their article *Lies, Damned Lies, & 400 000 Smoking-Related Deaths*).

This was followed by a report from the American Council on Science and Health (ACSH) which refuted each of the key arguments put forward by Levy and Marimont and supported the CDC estimate of 400 000 deaths per year.

Study the Levy and Marimont article and the ACSH report. Summarise the main points of difference. Which of the two estimates do you think is likely to be more accurate and why?

Lies, Damned Lies, & 400 000 Smoking-Related Deaths was published by the Cato Institute in *Regulation Volume 24* (1998).

You can look at these sites on the Internet by downloading these reports from the organisations' respective websites.

There are many other situations where different groups interpret statistical data in different ways or where one group gives more weight to one set of data while another group says a different set of data is more important.

For example, some people have concerns about the safety of the MMR vaccine, whereas other groups say there is no evidence of any problem. Opinions also differ about whether or not global warming is occurring and, if it is, whether the activities of humans are a significant cause.

MMR is the mumps, measles and rubella vaccination.

In most cases it is sensible to review all the available evidence before coming to any conclusions.

Activity 5.1G

Different views on the safety of the MMR vaccine
After carrying out an investigation into a possible link between autism and bowel problems, Andrew Wakefield of the Inflammatory Bowel Disease Study Group at the Royal Free Hospital said about the MMR vaccine, *'There is sufficient concern in my own mind for a case to be made for the vaccines to be given individually at not less than one year intervals'*.

The Government's view is that, on the scientific evidence available, there is no causal link between MMR vaccine and autism.

This view is also supported by the Medical Research Council and the World Health Organisation (WHO).

Find out as much as you can about research investigating the MMR vaccine. Write a summary of what you find.

The Government's committees of independent experts are:

- the Committee on Safety of Medicines (CSM) and
- the Joint Committee on Vaccination and Immunisation (JCVI).

Different views on global warming
The Intergovernmental Panel on Climate Change (IPCC) states in its Summary for Policy Makers *Climate Change 2001 The Scientific Basis* that *'Emissions of greenhouse gases and aerosols due to human activities continue to alter the atmosphere in ways that are expected to affect the climate'*. A comprehensive summary of the scientific work done to investigate global warming is given in a technical summary that accompanies this report.

Visit the IPCC and ABD websites to find the information you need.

In contrast, the Association of British Drivers (ABD) states in its article *Climate Change Truths* that *'There is no convincing scientific evidence for man-made global warming outside of computer models, which do not agree with actual climate data'*. The ABD provides much statistical evidence to back up this statement.

Compare and contrast the evidence used and conclusions reached by these two organisations.

5.2 Preparing for assessment

Your coursework portfolio

You have to include a report in your coursework portfolio in which you critically analyse the statistical work of others. This does not mean that you need to find lists of mistakes made in a report. As you will have seen when working through the activities of this chapter, there are often many questions to which you do not have answers when you read a report that incorporates statistics. You should highlight these. It may be particularly instructive to gather the reporting of some data by one or two sources, such as newspapers, and contrast these with the official data on which these reports are based.

You should consider critically what the report tells you about each part of the process of using and applying statistics to make sense of situations (see diagram).

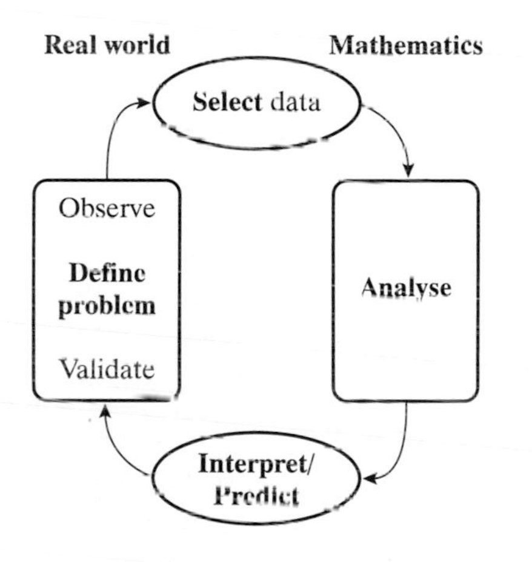

The Nuffield FSMQ website provides many data sets and web links that may be of use to you.

Remember to refer back to the questions raised at the start of this chapter on page 120.

6 Extension Opportunities

Contents

Using and Applying Statistics was designed to be a qualification of use to a wide range of students. Hopefully, you will find it useful in supporting your work towards your study of other subjects. You should be able to produce at least part of your coursework portfolio working with data you collect in another subject. The *What you need to learn* section of the specification lists a 'core' of statistical techniques and understanding that will be generally useful to all students, whatever their other studies and interests. The questions in the written examination will be set on this core content and the previous chapters in this book have covered this major part of the course.

However, there are many other statistical techniques that are likely to be useful to some students but not others. You are expected to develop competence in some areas that are beyond the core content and include some work relating to this in your coursework portfolio. Some possible areas of study are given in the specification as *Extension Opportunities* (listed in the margin here) but this is not intended to be limiting. There may well be other statistical techniques that are more relevant to your other work or interests and that you would rather use in your coursework portfolio.

This chapter is designed to give you some idea of what each of the extension opportunities listed in the specification involves. In each section a brief explanation of the technique is followed by one or more activities that will give you some practice in its use. You may decide to study one or more of these sections, possibly taking the subject much further than in this book.

For example, if you are studying A level Psychology or Biology, you may study the section on significance tests and then go on to study and use those tests in your subject to even greater depth. An assignment from A level Psychology or Biology might then form the basis for one of your investigations in your *Using and Applying Statistics* coursework portfolio.

On the other hand, if you are on an engineering or manufacturing course you may prefer to extend your knowledge of the control charts mentioned in the introduction to **Chapter 3** and then use this knowledge in your coursework portfolio. In this case you may decide to omit this part of the book altogether.

Note: the Nuffield FSMQ website provides some resources that you may find helpful.

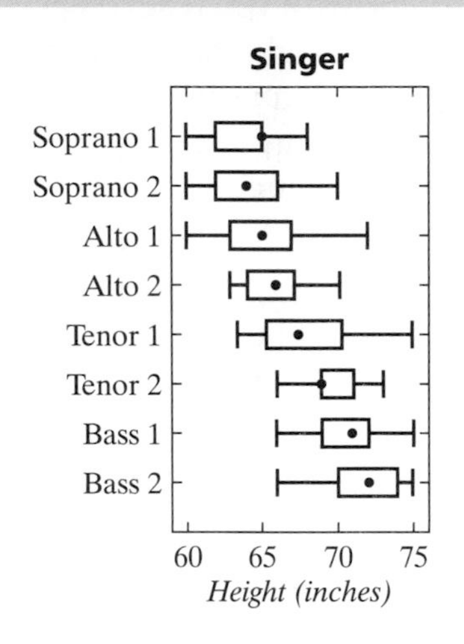

Extension Opportunities

Exploring, describing and comparing data sets using other diagrams appropriate to your own work such as back-to-back stem and leaf diagrams and box and whisker plots.

Other probability distributions that you may wish to use to model populations include: uniform, binomial, Poisson.

Use of rank correlation coefficients.
Use of non-linear regression models.

Significance tests that you could use:
t-test, Z-test, Mann Whitney U-test, Wilcoxon signed rank; Chi-square (χ^2).

6.1 Stem and leaf diagrams

In **stem and leaf diagrams** the data values themselves are organised in a way that shows the shape of the distribution.

For example, a group of female students were asked to record how long they spent on a computer during one week. The results in minutes are given alongside.

These data are shown as a stem and leaf diagram below.

Female students

130	184	216	75	150
215	160	506	240	260
565	80	138	205	125
321	164	280	415	90
132	145	324	634	458
346	350	82	136	401

Time spent on computers during a week by a group of female students

	Females									
0	7	8	8	9						
1	2	3	3	3	3	4	5	6	6	8
2	0	1	1	4	6	8				
3	2	2	4	5						
4	0	1	5							
5	0	6								
6	3									

Unit = 10 minutes

The **most significant part** of the data values are usually used as 'stems' and the **next most significant part** as 'leaves'.

In this sample the hundreds digits vary from 0 (when the value is less than 100) to 6. These digits are used as 'stems' and arranged vertically in order of size.

The tens digits are used as 'leaves'. Taking each of the values in turn, an entry corresponding to the tens digit is made in the table. The units digit in each value is ignored – i.e. the values are truncated.

Ignoring the units digit is equivalent to truncating all the values to the nearest 10 below.

Stem	Leaves		
0	7		This entry is from the fourth value of 75 (truncated to 070).
1	3	8	These entries are from the first two values, 130 and 184.
2	1		This entry is from the third value of 216 (truncated to 210).
3			
4			
5			
6			

Activity 6.1A

Write a few sentences describing the distribution of the time spent by female students on computers during this week

The results from a similar survey carried out with male students are given alongside.

A '**back-to-back**' stem and leaf diagram can be used to compare these results with those from the female students.

Male students

215	350	128	237	315
630	485	320	190	275
180	238	375	256	75
363	425	95	180	254
200	394	330	105	316
362	262	392	440	350

The stem and leaf diagram showing the time spent by female students on computers is shown again below. Note that the leaves in each row are arranged in order of size.

A stem and leaf diagram for male students has been started at the other side of the stem column to give a back-to-back diagram.

Discussion point
What advantages are there in arranging the leaves in order of size?

Time spent on computers during a week by a group of students

Males		Females									
	0	7	8	8	9						
	1	2	3	3	3	3	4	5	6	6	8
1	**2**	0	1	1	4	6	8				
5	**3**	2	2	4	5						
	4	0	1	5							
	5	0	6								
	6	3									

Unit = 10 minutes

Note the information that each unit is 10 minutes is given on the diagram. This means that anyone using the diagram can write down estimates of the original data values if they wish. For example, the final entry of 6|3 at the female side is equivalent to 630 minutes.

Discussion point
Why is it only possible to *estimate* the original data values from the diagram?

Activity 6.1B

1 **a** Complete the back-to-back stem and leaf diagram started above.

b Use your diagram to help you write a few sentences in which you compare the two distributions.

2 The scores of the students in an IQ test are given below:

Females

105	97	101	103	110	95	98	100	102	124
96	94	87	114	98	104	116	99	94	93
85	113	107	84	99	100	103	98	94	107

Males

113	109	115	102	95	96	98	93	104	108
93	99	106	98	114	100	97	104	110	92
99	87	100	105	96	93	104	90	113	82

Draw a back-to-back stem and leaf diagram to illustrate these scores. Use it to compare the IQ scores of the male and female students. Write a sesntence or two explaining your findings.

Discussion point
In this case it is more sensible to use 8, 9, 10, 11 and 12 as the stem values rather than the hundreds digits. Why?

6.2 Box plots

A **box plot** is a simple but effective way of illustrating a dataset. The sketch shows what a box plot looks like.

Values of the variable are shown on a horizontal or vertical axis. When using a vertical axis, the central bar stretches from the lower quartile to the upper quartile and has a line across it to show the median. The line at the top of the box plot joins the mid-point of the upper end of the bar to a point showing the maximum data value. The line at the bottom of the box plot joins the mid-point of the lower end of the bar to a point at the minimum data value. These lines are sometimes called 'whiskers' and the diagram is often called a '**box and whisker diagram**'.

It is useful to think of the box plot as having four parts, each part representing a quarter of the sample. The relative lengths of these parts can tell you a lot about the distribution.

In this case the whisker at the bottom is the longest part of the box plot. This means that the values in the bottom quarter of the distribution are more spread out than they are in any other part. The bottom section of the bar from the lower quartile to the median is shorter than the lower whisker. The values in this quarter of the distribution are less spread out than in the bottom quarter. The upper section of the bar and upper whisker are both shorter than the lower parts of the box plots. This indicates that values in the upper half of the distribution are closer together. Finally, the fact that the upper section of the bar and the upper whisker are approximately the same length suggests that the upper half of the distribution is fairly evenly spread.

You can deduce from the box plot what a histogram of the distribution might look like. The LQ, median and UQ divide the area below the histogram into four equal parts as shown below. This histogram gives a *rough* idea of the shape of the distribution, but because it is not known how the values are distributed within each part of the box plot it may not be very accurate.

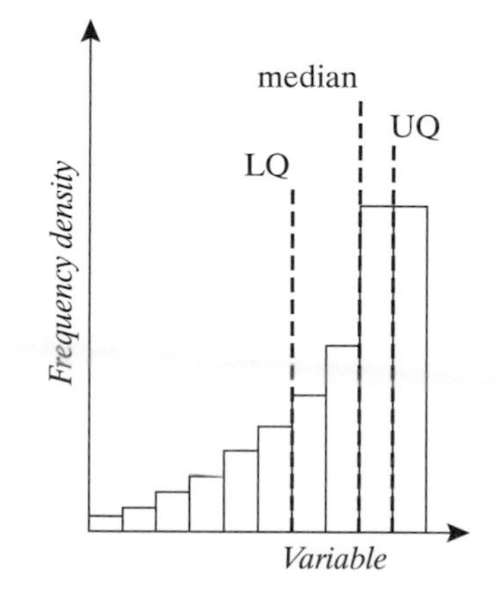

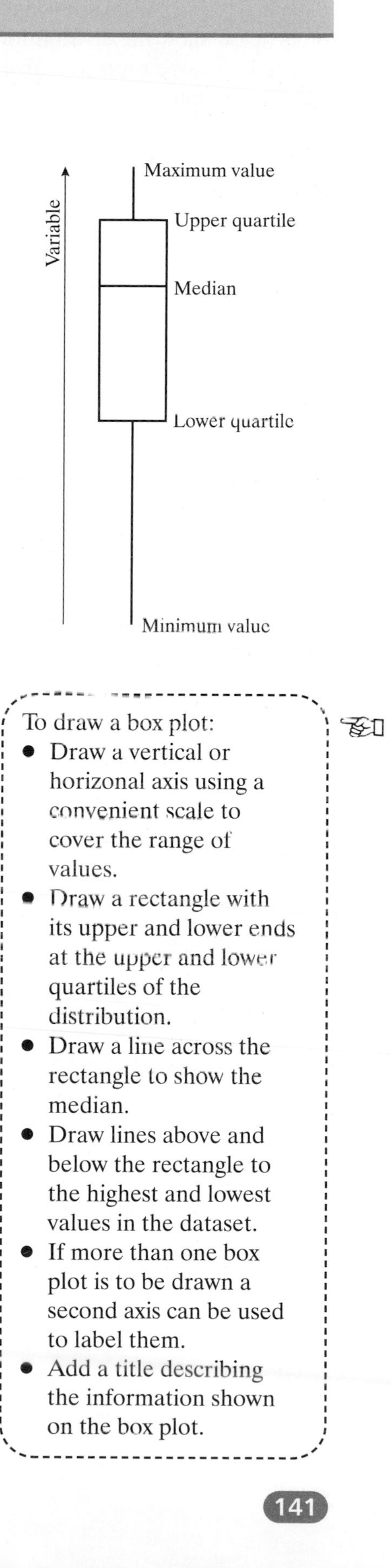

To draw a box plot:

- Draw a vertical or horizonal axis using a convenient scale to cover the range of values.
- Draw a rectangle with its upper and lower ends at the upper and lower quartiles of the distribution.
- Draw a line across the rectangle to show the median.
- Draw lines above and below the rectangle to the highest and lowest values in the dataset.
- If more than one box plot is to be drawn a second axis can be used to label them.
- Add a title describing the information shown on the box plot.

Box plots provide a good way of comparing distributions. The box plots below show the distribution of weight in large representative samples of adult men and women in the UK.

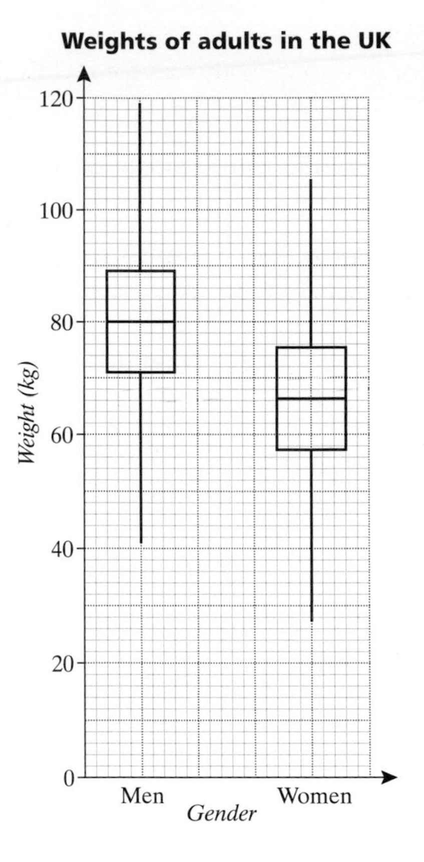

You may be able to use spreadsheet software to draw box plots, but sometimes it is necessary to alter slightly the way in which the median is shown. In the diagram below the medians are shown as crosses.

This diagram illustrates the results of a survey of the speeds of a large number of vehicles travelling along a stretch of motorway.

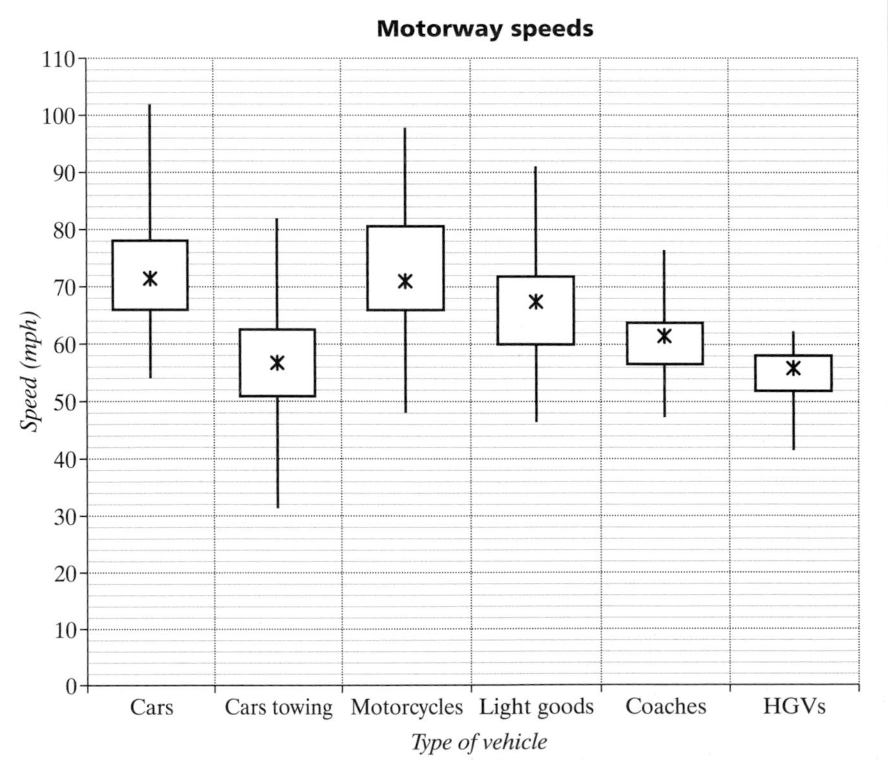

Discussion points

Make sure that you agree with the following interpretations of the box plots.

Comparison of the medians shows that on average the men weighed more than the women.

In each box plot the median line is approximately at the centre of the bar and the upper and lower whiskers are approximately equal in length. This symmetry in the box plots indicates that each sample is approximately symmetrical about the median value.

The fact that the whiskers are longer than the distance between the centre and ends of the bar suggests that the values at each end of the distribution are more spread out than they are near the centre of the distribution.

The lengths of the sections in the two box plots are similar. This suggests that the standard deviations of the two distributions are approximately equal.

Usually weights follow a normal distribution. All the points made above are consistent with this and the usual shape of the histogram of a normal distribution (shown below).

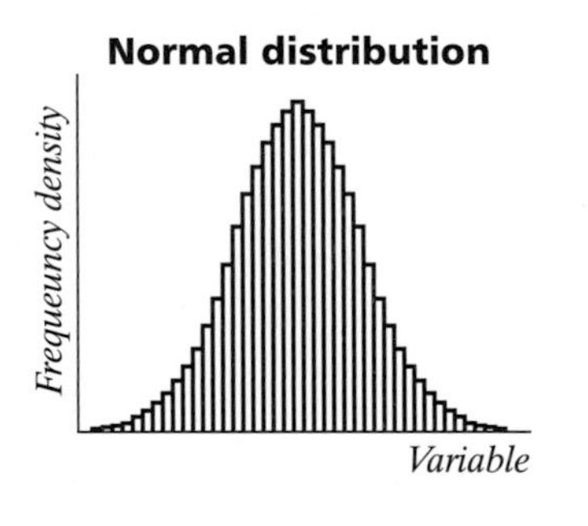

HGVs are heavy goods vehicles.

The speed limit for cars, motorcycles, light goods vehicles and coaches on the motorway is 70 miles per hour, while that for HGVs and cars towing something is 60 miles per hour.

Discussion point

Do drivers stick to the speed limits?

Activity 6.2A

1 a Use the weight box plots to complete the following table:

Gender	Median	Min	Max	Range	LQ	UQ	IQR
Men							
Women							

b Use the results in the table to compare and contrast the weight distributions of the men and women. Compare your comments with what was said in the discussion points.

2 Use the speed box plots to write a brief report describing the extent to which the drivers of different types of vehicles disobey the motorway speed limits.

3 a Use the speed box plots to describe the distribution of motorway speeds for

i cars

ii motorcycles

iii HGVs.

b For each of these speed distributions sketch the likely shape of a histogram.

4 The histograms below show different shapes of distributions. In each case draw a sketch to show what you think the corresponding box plot would look like.

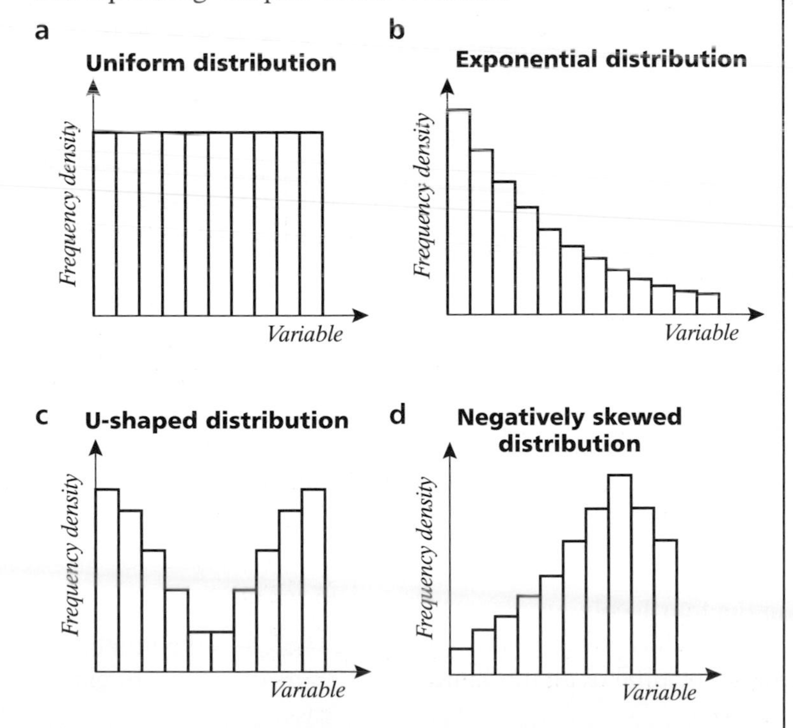

6.3 More about correlation and regression

Spearman's rank correlation coefficient

In **Chapter 2** you saw that correlation coefficients measure the degree of association between two variables.

Pearson's correlation coefficient uses pairs of data assumed to be drawn from normal distributions and to have a linear connection.

By contrast, **Spearman's rank correlation coefficient** measures the degree of association between the orders in which objects are placed for two different properties.

Consider the results of ten students who took a two-section exam. The results are given as orders of merit. The table shows their positions (ranks) in each of the two sections. For example, Student A came second in Paper 1 and first in Paper 2.

Student	Paper 1 rank	Paper 2 rank
A	2	1
B	8	7
C	1	3
D	3	2
E	7	10
F	9	8
G	5	5
H	4	4
I	6	6
J	10	9

Note: If two or more students have the same rank for either of the sections, for example if the ranks are 1, 2, 2, 4, …., the 'tied' students are given a mean rank, so in this case the ranks used are 1, 2.5, 2.5, 4, ….

You can see that, not surprisingly, students did not always gain the same rank in each section, but the ranks for the two sections are fairly similar. You do not have the actual marks the students achieved on each paper so you cannot calculate Pearson's product moment correlation coefficient. In this case you can calculate Spearman's rank correlation coefficient. This measures the strength of the relationship between the two sets of ranked data.

Discussion point

Why would you expect close, but not total, agreement in the rankings in this case?

For the above data you calculate the difference between the two ranks for each student by subtracting the rank for Paper 2 from the rank for Paper 1 and write the results in the fourth column.

Student	Paper 1 rank	Paper 2 rank	Difference in rank, d	d^2
A	2	1	1	1
B	8	7	1	1
C	1	3	−2	4
D	3	2	1	1
E	7	10	−3	9
F	9	8	1	1
G	5	5	0	0
H	4	4	0	0
I	6	6	0	0
J	10	9	1	1
				$\Sigma d^2 = 18$

Discussion point

Explain why the total of the numbers in the fourth column will always be zero. Where have you met this method of squaring to get positive values before?

The bigger the differences in the ranks, the less strong is the relationship.

If you find the total of the numbers in the fourth column, you will get zero, so first you square the values of d, which means there will now be no negative values, then calculate the total of all the values in the d^2 column.

The formula used to calculate Spearman's rank correlation coefficient is

$$r_s = \frac{1 - 6\Sigma d^2}{n(n^2 - 1)},$$

where n is the number of pairs of values, 10 in this case.

Activity 6.3A

You will find it useful to work with a spreadsheet when calculating values in this activity.

1 Show that Spearman's rank correlation coefficient, r_s, for this data is 0.89. What sort of relationship does this indicate?

2 Find the values of d^2 for

a perfect rank agreement

b completely opposite ranks for the two sections of the test. Hence show that the maximum and minimum possible values of r_s are 1 and -1.

3 The data in the table on the right show information about GCSE results and numbers of pupils having free school meals in 14 schools – you have used such data before, but this time the schools are ranked.
Calculate Spearman's rank correlation coefficient for the data in the table and interpret what your result tells you.

School	Rank for GCSE results	Rank for percentage of free school meals
A	8	13
B	12	2
C	11	7
D	13	6
E	9	14
F	7	4
G	6	3
H	2	11
I	14	1
J	5	8
K	4	9
L	3	12
M	1	10
N	10	5

4 Twenty of the richest countries in the world are ranked in terms of Gross Domestic Product (GDP) per person and economic growth rate.

a What is the strength of the relationship between GDP per person and economic growth rate as indicated by the information in this table?

Country	GDP per person rank	Growth rate rank
Australia	16	3
Austria	9	11
Belgium	6	13
Canada	11	7
Denmark	7	16
Finland	19	8
France	12	10
Germany	14	14
Iceland	8	2
Italy	18	17
Japan	10	20
Kuwait	15	18
Luxembourg	1	4
Netherlands	13	9
Norway	5	19
Singapore	3	1
Sweden	20	6
Switzerland	4	15
United Kingdom	17	12
United States	2	5

b Calculate Spearman's correlation coefficient for the ranks in this table and interpret your result.

Non-linear correlation and regression

Consider the data for a cooling cup of coffee shown in the graph alongside.

Just as you can use the method of least squares to find a linear regression line, the method of least squares can also be applied to finding, for example, quadratic or exponential regression lines and the corresponding correlation coefficients. Of course, the calculations are more complicated for more complex functions!

Spreadsheet programs will draw 'trend lincs', which can be polynomial, logarithmic, or exponential as well as linear. The equations of such lines are determined using the method of least squares.

You may like to make use of such regression lines in your work where appropriate.

Cooling cup of coffee

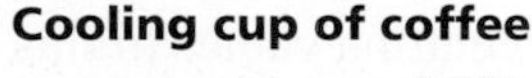

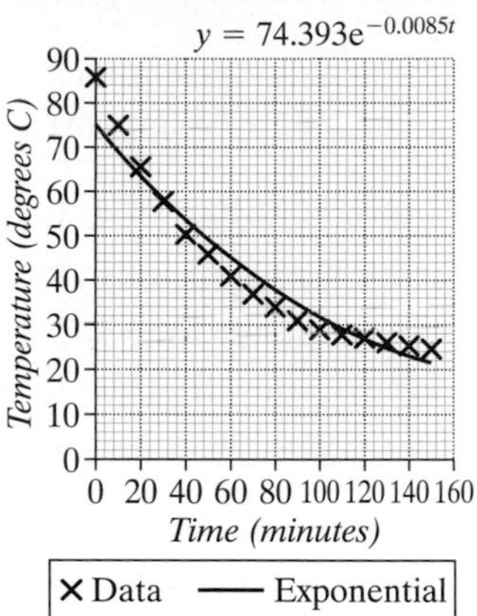

A spreadsheet program gives the function $y = 74.4e^{-0.0085x}$ as an exponential to fit this data.

Activity 6.3B

The table below shows the length and the mass of members of the cat family.

Use a spreadsheet program to plot a scatter graph of the data. Investigate what type of trend line (linear, quadratic, cubic, exponential, etc.) best fits the data.

Type of cat	Length (cm)	Mass (kg)
Bobcat	70	11
Cheetah	150	65
Domestic cat	50	5
Jaguar	180	180
Leopard	150	91
Lion	240	227
Lynx	110	18
Ocelot	100	16
Puma	180	91
Tiger	270	270

6.4 Probability distributions

The uniform distribution

In uniform distributions a random value of the variable is *equally likely* to be any of the possible values.

Random digits from tables (or a calculator or computer) give a *discrete* uniform distribution. The only possible values of the random digits are 0, 1, 2, 3, 4, 5, 6, 7, 8 and 9, each having a probability of $\frac{1}{10}$. The bar chart illustrates this distribution.

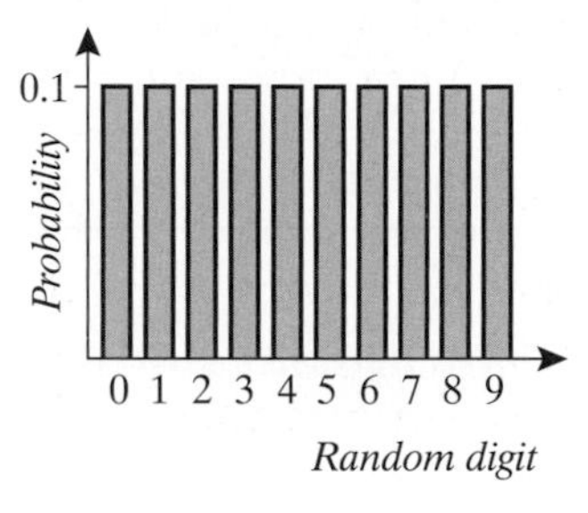

In a *continuous* uniform distribution each value in a *range* of values is equally likely to occur. If this range extends from a minimum value of a to a maximum value of b, the probability of any particular value occurring at random is given by the probability density function (pdf): $f(x) = \frac{1}{b-a}$

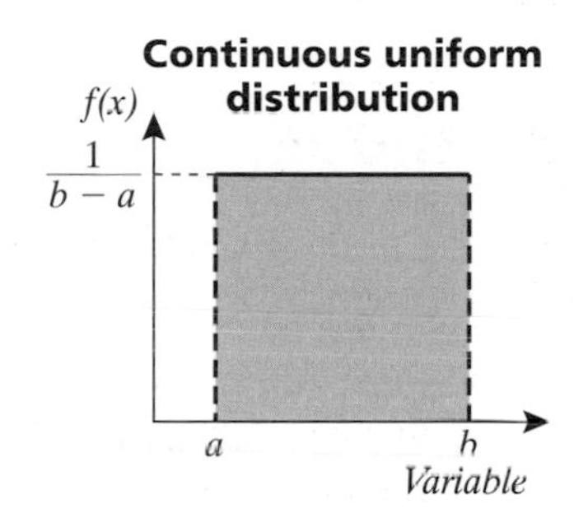

The area under the graph of this function is 1. Because this area is in the shape of a rectangle the distribution is often called the **rectangular distribution**.

The theoretical mean and standard deviation of a continuous uniform distribution are given by:

$$\mu = \frac{a+b}{2} \text{ and } \sigma = \sqrt{\frac{(b-a)^2}{12}}$$

Discussion point

Can you explain why the mean is $\mu = \frac{a+b}{2}$?

Activity 6.4A

1 a Explain why the score on one throw of a dice follows a uniform discrete distribution but the total score when two dice are thrown does not.

b Draw a bar chart to illustrate the distribution of the score on one throw of a dice.

2 When the length of the components from a production line are recorded to the nearest 5 millimetres the rounding errors are uniformly distributed in the range from −2.5 mm to 2.5 mm.

a Draw a sketch of the distribution.

b Find

i the probability density function of this distribution

ii the theoretical mean

iii the theoretical standard deviation.

3 Before it is spun the arrow on a wheel of fortune is always in the position shown by OA. When it stops the arrow is in the position OB. A and B lie on a circle of radius 1.2 metres. Assume that B is equally likely to lie at any point on the circumference of this circle.

a Describe the distribution of
  i the smaller angle AOB **ii** the minor arc length AB.

b Draw a sketch of each of these distributions and in each case show on your sketch the pdf and the mean value.

Wheel of Fortune

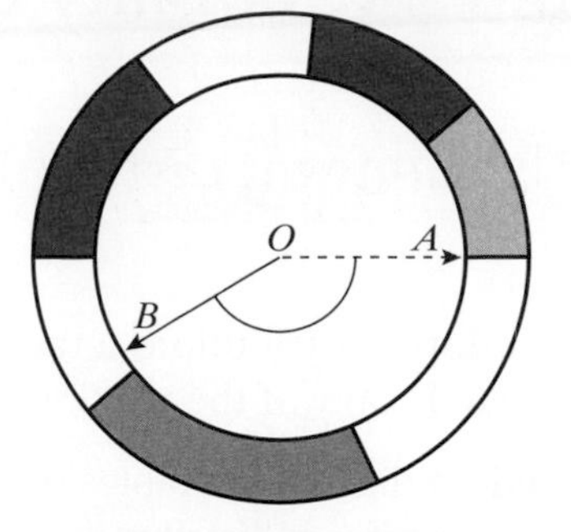

The binomial distribution

A **binomial distribution** arises when an 'experiment' is repeated a number of times and at each attempt there are only two possible outcomes, usually called 'success' and 'failure', and where the probability of success at each attempt remains constant.

Note that the 'experiment' can be any activity that is carried out a number of times and 'success' may not always mean something good.
The trials must be **independent**, i.e. the outcome of each trial must not affect the outcome of any other trial.

For example, you might toss a dice five times and count each six you get as a success. If the dice is not biased the probability of success at each of the attempts is $\frac{1}{6}$ and the probability of failure is $\frac{5}{6}$.

The **parameters** of a binomial distribution are n, the number of attempts and p, the probability of success at a single attempt. The random variable is X, the number of successes from the n attempts. The probability of each possible value of X, from 0 (no successes) to n (all successes), can be found from the values of n and p.

Parameters are constants that define the distribution.

Note that the binomial distribution is **discrete** since all possible values of the variable are whole numbers.

In the case where you throw a dice five times $n = 5$ and $p = \frac{1}{6}$.
The probabilities of 0, 1, 2, 3, 4, 5 successes occurring are:

$$P(X = 0) = \left(\frac{5}{6}\right)^5 = 0.402$$

$$P(X = 1) = 5 \times \frac{1}{6} \times \left(\frac{5}{6}\right)^4 = 0.402$$

$$P(X = 2) = 10 \times \left(\frac{1}{6}\right)^2 \times \left(\frac{5}{6}\right)^3 = 0.161$$

$$P(X = 3) = 10 \times \left(\frac{1}{6}\right)^3 \times \left(\frac{5}{6}\right)^2 = 0.0322$$

$$P(X = 4) = 5 \times \left(\frac{1}{6}\right)^4 \times \frac{5}{6} = 0.003\,22$$

$$P(X = 5) = \left(\frac{1}{6}\right)^5 = 0.000\,129$$

P(0 successes) = P(5 failures)
= P(F, F, F, F, F)
= P(F) × P(F) × P(F) × P(F) × P(F)

P(1 success and 4 failures)
= P(S, F, F, F, F)
= 5 × P(S) × P(F) × P(F) × P(F) × P(F)
The probability is multiplied by five because there are five different ways in which the success and failures could occur. (Can you list them?)

P(2 successes and 3 failures)
= P(S, S, F, F, F)
= 10 × P(S) × P(S) × P(F) × P(F) × P(F)
The probability is multiplied by ten because there are ten different ways that 2 successes and 3 failures could occur. (Can you list these?) Can you explain the other probabilities?

Note the pattern in the form of these calculations.
This is a feature of all binomial distributions.

Since these are all the possible results when a dice is thrown five times, the total probability should be 1. Check whether this is so.

If the number of attempts, n, is large it is much more difficult to find the number of ways each outcome can occur. Luckily there is a formula that gives binomial probabilities in terms of n and p:

$$P(X = x) = \frac{n!}{x!(n-x)!} p^x (1-p)^{n-x} \text{ for } x = 0, 1, 2, \ldots, n$$

number of different ways — x successes — $n - x$ failures

$n!$ is called **factorial n** and means:
$n \times (n-1) \times (n-2) \times \ldots \times 3 \times 2 \times 1$
e.g. $5! = 5 \times 4 \times 3 \times 2 \times 1 = 120$
Find out how to use your calculator to find factorials and check the value of 5!

Check that this formula gives the probabilities listed above.

The binomial distribution is useful in many real situations. The following activity involves a few of these.

Activity 6.4B

1 A market gardener keeps records of the germination rates of the peas he plants. His records show that when he plants a pea the probability that it germinates is $\frac{4}{5}$.
If he plants a row of ten peas find the probability that

a they all germinate
b none of them germinate
c half of them germinate
d eight or more germinate.

2 A maternity hospital's records show that the probability of a baby being a boy is 0.51. Use the binomial distribution to estimate the probability that of the next twelve babies born at the hospital

a exactly half will be boys
b more than ten will be boys.

3 A manufacturer of wine glasses has calculated that 5% of the glasses produced are broken during transit to sales outlets.

a A shop has ordered 20 glasses.
Calculate the probability that
i all the glasses will survive the journey to the shop intact
ii more than three glasses will be broken on arrival.

b The manufacturer packs the glasses into standard packs before transit. The number of glasses in each standard pack is given below:

Pack A 20 glasses — Pack B 50 glasses
Pack C 100 glasses — Pack D 500 glasses

Find the mean number of broken glasses for each standard pack.

4 A driving instructor has found that on average three out of every five learners he enters for a driving test pass.

a He uses the binomial distribution to estimate the probability that eight or more of his next ten pupils taking the test will pass. Calculate his result.
b Why is his result not likely to be accurate?

The theoretical mean and standard deviation of a binomial distribution can also be given in terms of n and p:
$\mu = np$ and $\sigma = \sqrt{np(1-p)}$

Here a 'success' can be defined as a broken glass!

Discussion point
Why is it easier to find the probability that three or less will be broken? How can you use this fact?

Discussion point
Which of the conditions for a binomial distribution are not likely to be satisfied in this case?

The Poisson distribution

The **Poisson distribution** models events that are randomly distributed over time or space, for example:

- telephone calls taken by a receptionist during an hour
- car accidents on a particular stretch of road during a year
- weeds appearing in an area of garden
- flaws in a length of fabric
- particles emitted by a radioactive source in a given interval of time.

In a Poisson distribution the events must occur singly, i.e. no more than one occurrence should happen at the same time or place. The conditions of the context should remain the same, e.g. in the case of car accidents there should be no change in road conditions.

Suppose the random variable, X, is the number of occurrences in a given interval of time or space and λ is the mean number of occurrences in that size of interval. Then the probability of x occurrences is given by the formula:

$$P(X = x) = \frac{e^{-\lambda}\lambda^{x}}{x}! \text{ for } x = 0, 1, 2, 3, \ldots.$$

The theoretical mean and standard deviation are given by:
$\mu = \lambda$ and $\sigma = \sqrt{\lambda}$

λ is the **parameter** of this distribution.

$x!$ is called **factorial x** and means:
$x \times (x - 1) \times (x - 2) \times \ldots \times 3 \times 2 \times 1$
e.g. $5! = 5 \times 4 \times 3 \times 2 \times 1 = 120$
If you have not met this before, find out how to use your calculator to find factorials and check the value of 5!
Note: 0! is defined to be 1.

If the mean is known for a particular interval, then its values can be deduced for other intervals.

For example, suppose that an insurance company receives on average 3 claims per day, then during a 6-day week it will receive on average 18 claims.

You can find the probability that the insurance company receives x claims on a particular day from $P(X = x) = \frac{e^{-3}3^{x}}{x!}$ and the probability that the insurance company receives x claims in a particular week from $P(X = x) = \frac{e^{-18}18^{x}}{x!}$.

Discussion point
What assumptions have been made to use the Poisson distribution in this context? Do you think they are valid?

The probability that the insurance company receives less than 3 claims on a particular *day* is:

$$P(X = 0) + P(X = 1) + P(X = 2) = \frac{e^{-3}3^{0}}{0!} + \frac{e^{-3}3^{1}}{1!} + \frac{e^{-3}3^{2}}{2!}$$
$$= e^{-3} + 3e^{-3} + \frac{9e^{-3}}{2}$$
$$= 8.5e^{-3} = 0.423 \text{ (to 3 d.p.)}$$

The probability that the insurance company receives less than 3 claims in a particular *week* is:

$$P(X = 0) + P(X = 1) + P(X = 2) = \frac{e^{-18}18^{0}}{0!} + \frac{e^{-18}18^{1}}{1!} + \frac{e^{-18}18^{2}}{2!}$$
$$= e^{-18} + 18e^{-18} + 162e^{-18}$$
$$= 181e^{-18} = 0.000\,0028 \text{ (to 7 d.p.)}$$

This is very small, as you would expect since the company normally has on average 3 claims per day.

Discussion point
What is the probability that the insurance company receives 3 or more claims on

- a particular day?
- a particular week?

Activity 6.4C

1 Telephone calls are received at a college office at a mean rate of 12 per hour between 10 am and 4 pm.

a Use a Poisson distribution to calculate the probability that in a particular hour

i no telephone calls are received

ii less than three telephone calls are received

iii three or more telephone calls are received.

b **i** What is the mean rate of telephone calls received ***per minute*** during the period from 10 am to 4 pm?

ii Use a Poisson distribution to calculate the probability that no telephone calls are received in a particular minute.

2 Cars arrive at a petrol station at an average rate of one per minute.

a Use a Poisson distribution to calculate the probability that in a particular minute

i no cars arrive at the petrol station

ii four cars arrive at the petrol station

iii less than four cars arrive at the petrol station.

b Use a Poisson distribution to calculate the probability that in a particular ***ten-minute*** interval

i no cars arrive at the petrol station

ii four cars arrive at the petrol station

iii less than four cars arrive at the petrol station.

c What assumptions have been made in using the Poisson distribution in this context? Do you think these assumptions are valid?

3 The ribbon produced by a factory has an average of two flaws per 100 metres.

a Assuming that the number of flaws per 100 metres follows a Poisson distribution, what is its standard deviation?

b Find the probability that a hundred metre length of ribbon has

i no flaws **ii** just one flaw

iii more than one flaw.

c The factory produces 5000 metres of ribbon per day. Find the probability that the ribbon produced by the factory on one day has less than 10 flaws.

d Find the probability that a one metre length of ribbon has

i no flaws **ii** at least one flaw.

Discussion point

Why should this method only be used to give probabilities for hours between 10 am and 4 pm and not for times before 10 am or after 4 pm?

6.5 Significance tests

The distribution of the sample mean

Significance tests are used to test hypotheses. In some cases this is done by testing the mean of a sample. Before looking at how such a test works we need to study the **distribution of the sample mean.**

Suppose samples of adult men are taken from some districts selected at random in the UK and the mean weight of each sample is calculated. The sample mean, $\bar{x}$, will vary from one sample to another, following a distribution that is known as the **distribution of the sample mean**. If the samples are large and random, then you can expect each value of the random variable, $\bar{X}$, to be quite close to the population mean μ for the whole population of adult men in the UK. Sometimes $\bar{x}$ will be a bit smaller than μ and sometimes a bit larger.

Discussion point

How will the size of the sample affect the variability of the sample mean?

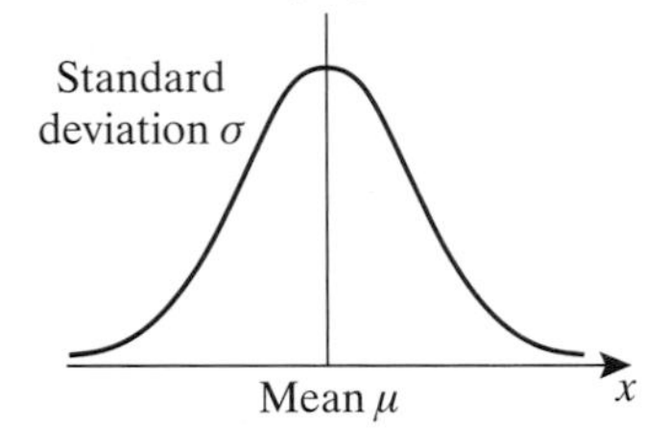

In general, if samples of size n are taken from a ***normal*** population with mean μ and standard deviation σ, then $\bar{X}$ follows a ***normal*** distribution with mean μ and standard deviation $\dfrac{\sigma}{\sqrt{n}}$.

The sketches show a normal distribution and the distribution of the mean of samples of size n from that distribution. The distribution of the sample mean has the same central value as the underlying distribution, but is less wide. The larger the number of items in the samples, the narrower the distribution of sample means.

Discussion point

Can you explain why the larger the number of items in the samples the narrower the distribution of sample means?

Even when the samples are from some other type of distribution $\bar{X}$ still follows an *approximately normal* distribution with mean μ and standard deviation $\dfrac{\sigma}{\sqrt{n}}$ if n is *large*. This is the **Central Limit Theorem** and $\dfrac{\sigma}{\sqrt{n}}$ is often called the **standard error** of the mean.

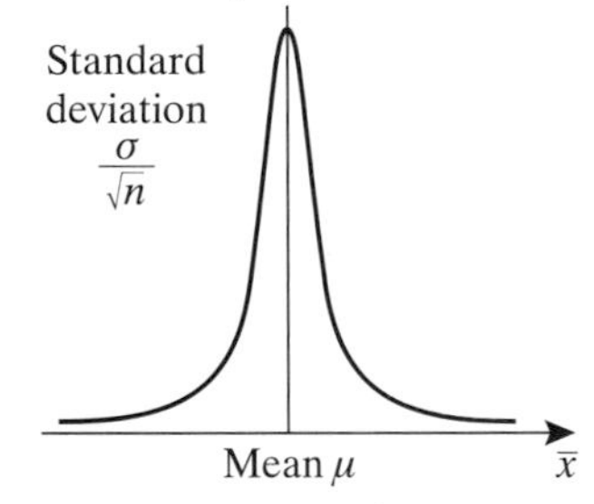

Activity 6.5A

For each of the following situations:

- **a** Draw a sketch of the distribution of the whole population.
- **b** Find the mean and standard deviation of the sample means.
- **c** Sketch the distribution of the sample means.

1 Samples of 400 adult women taken from the UK population whose heights vary normally with mean 1620 mm and standard deviation 64.4 mm.

2 Samples of size 50 taken from the production line of a biscuit factory where the weights of the biscuits produced vary normally with mean 32 g and standard deviation 1.5 g.

3 Samples of 1000 electrical components with lifetimes from a normal distribution with mean 5000 hours and standard deviation 450 hours.

Normal significance test

Suppose a market gardener knows that the yield she gets from a particular type of tomato plant is normally distributed with mean 3.24 kilograms (per plant) and standard deviation 0.56 kilograms. She wants to know whether the yield will be improved if she changes the type of fertiliser she uses. To answer this question she is likely to carry out an experiment in which she uses the new fertiliser on a number of plants and measures the yield she gets.

Suppose she uses the new fertiliser on 100 plants and finds that the mean yield from these plants is 3.38 kilograms. This is bigger than the population mean was before she used the new fertiliser, but not that much bigger. Perhaps this result has just occurred because of the inherent variability between samples of a distribution. Perhaps the yield from these particular plants would have been as good with the old fertiliser. How can she decide whether this result is good enough to suggest that the new fertiliser is better than the old one?

The market gardener can use a z test to help her decide whether the yield after using the new fertiliser is **significantly** better than it was before. The working below shows how to do this.

First state a **null hypothesis** and an **alternative hypothesis**:

Null Hypothesis, H_0: $\mu = 3.24$ kg
Alternative Hypothesis, H_1: $\mu > 3.24$ kg

The **null hypothesis**, denoted by H_0, assumes that there has been *no change* in the population mean, i.e. that the new fertiliser has had no effect. The **alternative hypothesis**, H_1, states that the mean is now *greater* than 3.24 kg, meaning that the fertiliser has *increased* the average yield.

On the basis of the null hypothesis, the distribution of the sample mean can now be identified:

$\overline{X}$ follows a normal distribution with mean $\mu = 3.24$ kg and standard deviation $\frac{\sigma}{\sqrt{n}} = \frac{0.56}{\sqrt{100}} = 0.056$ kg.

Now find the corresponding Z value (called the **test statistic**).

Using the distribution of the sample mean $Z = \frac{X - \mu}{\sigma}$ becomes

$$Z = \frac{\overline{X} - \mu}{\frac{\sigma}{\sqrt{n}}}$$

The test statistic in this case is $z = \frac{3.38 - 3.24}{0.056} = 2.5$.

Note that the value from the experiment is used as $\overline{X}$.

In order to complete the test we now need to select a range of z values that tells us when we should reject the null hypothesis. This set of z values is called the **critical region** and depends on the **significance level** of the test.

The sketch below shows the situation for a significance test at the 5% level.

The **critical value** of z for a 5% significance test is found from the *Standard Normal Table* by looking up the probability of 0.95. Check that the table gives the value shown in the sketch, i.e. 1.645.

5% of the distribution lies in the upper tail where $z > 1.65$.
This is the **critical region** for a 5% upper-tailed significance test and $z = 1.65$ is the **critical value** of z.

If the null hypothesis is true then only one in every 20 random values of the sample mean lies in this critical region. The fact that the test statistic, $z = 2.5$, *is in the critical region* means that the value obtained in the experiment is *very unusual*. The null hypothesis should be rejected. This is said to be a **significant** result at the 5% level.

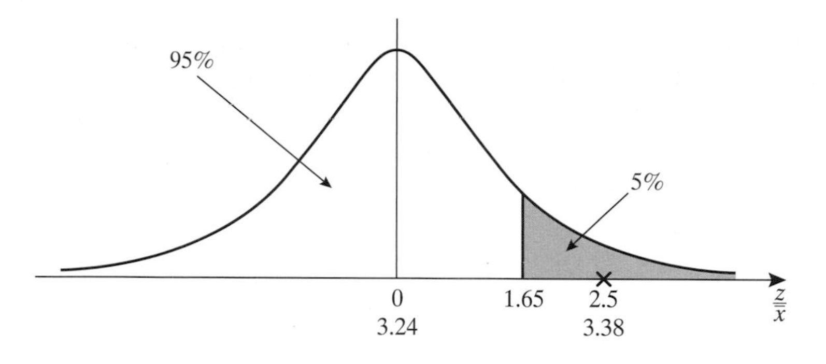

The **critical region** is a range of unusual values. How unusual depends on the significance level of the test. **When the test statistic lies in the critical region the null hypothesis should be rejected.**

The conclusion that should be drawn from this test is that the new fertiliser has given an increased yield of tomatoes.

The result of a significance test should always be given in terms of the real situation.

Many of the significance tests carried out in real situations are at the 5% level, but sometimes other levels are used. If the researcher or experimenter wants to be even more certain about the conclusion then a 1% significance test can be used instead. The critical region for a 1% upper-tailed significance test is $z > 2.33$.

In the example above the test statistic of 2.5 is in this critical region and the null hypothesis would also be rejected at the 1% level.

Sometimes an experiment might be used to test whether the mean has been *reduced* rather than increased. In this case the critical region would be in the *lower tail*.

The critical region for a 5% lower-tailed significance test is $z < -1.65$ and for a 1% lower-tailed significance test $z < -2.33$.

Discussion points

How was the value of 2.33 found from the *Standard Normal Table*?

What would be the critical region for a 10% upper-tailed test?

What do these values mean in terms of probability?

Activity 6.5B

1 The climbing ropes made by a manufacturer have breaking strengths that follow a normal distribution with mean 176.4 kg and standard deviation 13.6 kg. The manufacturer replaces one of the materials used in the rope with another. He tests a sample of 64 ropes and finds that their mean breaking strength is 180.2 kg.

- **a** Use a 5% significance test to decide whether the new material has increased the breaking strength of the ropes.
- **b** Would the conclusion be the same if a 1% significance test was used?
- **c** Explain what your answers to **parts a** and **b** tell you about the situation.

2 The heights of a variety of tulips grown at a nursery are normally distributed with mean 31.9 cm and standard deviation 1.25 cm. An experiment is carried out to find out whether the height is increased if the tulips are watered more frequently. A sample of 25 tulips is watered more frequently. These tulips grow to a mean height of 32.3 cm.
Carry out a 5% significance test to decide whether the extra watering has led to an increase in height.

3 The number of errors per page of text made by a typist has a mean of 6.4 and standard deviation 1.9. After a training course the typist is tested on 36 pages of text. He makes a total of 196 errors in this test.

a Carry out a 5% significance test to decide whether the training course has improved his performance.

b Would the conclusion be the same if a 1% significance test was used?

Discussion point

What assumptions do you need to make to carry out a *z* test in this context?

4 A packet of breakfast cereal says it contains 500 g, but the manufacturer claims that the mass of cereal in the packs is in fact normally distributed with mean 503 g and standard deviation 1.5 g. An inspector tests the manufacturer's claim by weighing the contents of 100 packets. The mean mass is found to be 501.8 g.

a Carry out a significance test at the 1% level to decide whether the manufacturer's claim is true.

b Also carry out a significance test at the 1% level to decide whether the population mean is greater than 500 g.

c Do you think the manufacturer is giving a fair deal? Explain your answer.

The previous examples have tested a rise or fall in the population mean using a **one-tailed** significance test. In cases where it is not known whether the true population mean may be above or below the value being tested, a **two-tailed** test should be used.

For example, suppose a machine is used to cut metal rods. The machine is set to cut lengths of 250 mm but the actual lengths cut by the machine follow a normal distribution with mean 250 mm and standard deviation 1.2 mm. After the machine is cleaned it is thought that the setting may need adjustment. To test this a sample of 25 rods are measured.

The mean length of these rods is 249.6 mm.

A 5% significance test can use this result to decide whether or not the machine needs adjustment.

Null Hypothesis, H_0: $\mu = 250$ mm
Alternative Hypothesis, H_1: $\mu \neq 250$ mm (2-tailed test)

Assuming H_0 is true, $\overline{X}$ follows a normal distribution with mean $\mu = 250$ mm and standard deviation $\dfrac{\sigma}{\sqrt{n}} = \dfrac{1.2}{\sqrt{25}} = 0.24$ mm.

Using $Z = \dfrac{\overline{X} - \mu}{\frac{\sigma}{\sqrt{n}}}$ the test statistic is $z = \dfrac{249.6 - 250}{0.24} = -1.\dot{6}$.

In a two-tailed 5% significance test the critical region is split into two parts, each part containing 2.5% of the distribution. Using the *Standard Normal Table* gives the critical region $z < -1.96$ and $z > 1.96$.

Use the *Standard Normal Table* to check this value of z.

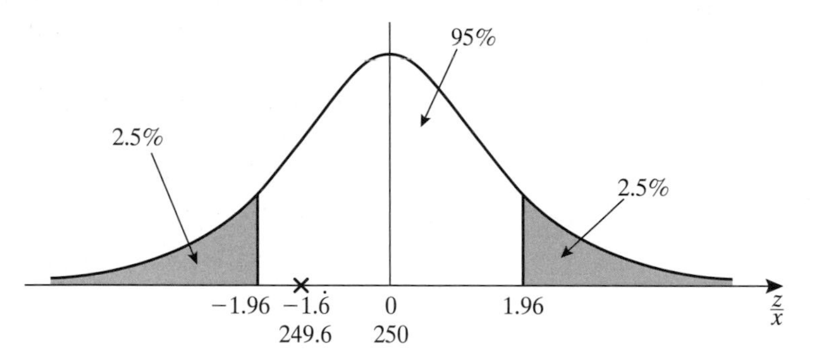

As the test statistic $z = -1.\dot{6}$ does not lie in the critical region H_0 is accepted. There is insufficient evidence that the machine should be adjusted.

Note that since $-1.\dot{6}$ is not far from -1.96 it may be prudent to carry out more tests in this case.

Activity 6.5C

1 The time spent by visitors at an exhibition is normally distributed with mean 72.1 minutes and standard deviation 23.2 minutes. After one item in the exhibition is exchanged for another, the time spent in the exhibition by a random sample of 100 visitors is measured. The mean time from this sample is 76.9 minutes.

a Carry out a 5% significance test to decide whether the exchange of items has led to a change in the mean time spent by visitors at the exhibition.

b Do you think this test would be valid if the sample of visitors used were simply the next 100 visitors after the item was replaced? Explain your answer.

2 The daily takings from a corner shop follows a normal distribution with mean £953 and standard deviation £62. The shopkeeper changes the way she displays her stock and finds that her total takings over the next 35 days is £32 486.

a Carry out a significance test at the 5% level to decide whether the change in the way in which the stock is displayed has affected takings.

b Would the result have been the same from a 1% significance test?

c What would you advise the shopkeeper to do?

Note that a z test can also be used to test the difference between two samples.

Note that the rest of the tests in this section require the use of other sets of tables and a full treatment is beyond the scope of this book. Brief descriptions are included to give you some idea of what each test can do and help you to decide whether further study of one or more of them would be useful to you.

The tests covered here are those mentioned in the specification for *AS: Use of Mathematics*. Other tests exist and can be used instead if you prefer.

Student's *t* test

When the standard deviation of the population, σ, is not known and the sample size is small (<30) it is not appropriate to use the normal distribution to test a value of the population mean. In such situations a **Student's *t* test** is used instead of a z test.

When the population variance is not known but the sample size is ***large*** the normal distribution ***can*** be used because the sample standard deviation can provide a good enough estimate of σ.

The method for using a t test to test the population mean is very similar to the method for a z test with the test statistic defined as:

$$T = \frac{\overline{X} - \mu}{\left(\frac{S}{\sqrt{n-1}}\right)}$$ where S is the sample standard deviation.

A t test can also be used to test whether there is a significant difference between two samples. Carrying out such tests requires the use of t tables. This is beyond the scope of this book.

Mann Whitney *U* test

The z and t tests are **parametric** tests. In some situations very little is known about the parameters of the distribution and other tests, called **non-parametric** tests, are used instead. The **Mann–Whitney test** is one of these. It tests whether there is a significant difference between two samples. For example, it could be used to test whether there is a significant difference between the performance of boys and girls when they perform a task.

Parametric tests are used to test the value of a parameter, often μ, of a distribution.
Non-parametric tests do not require any knowledge of the parameters.

The null hypothesis is that the distributions from which the two samples come are identical and the alternative hypothesis is that they are not the same.

In the test the two samples, A and B, are combined and ranked in order of size. If there is a significant difference between the two samples then you would expect the readings from one of the samples to be generally larger than those from the other sample. In this case the readings from one sample will be concentrated at one end of the ordered list and those from the other sample at the other end.

The test statistic, U, is calculated as follows:

- Each value in sample A is given a 'point' for each value in sample B that it exceeds and the total number of points for sample A is calculated.
- Each value in sample B is given a 'point' for each value in sample A that it exceeds and the total number of points for sample B is calculated.
- The smaller of these two sums is taken as the test statistic U.

The value of U is then looked up in a table to determine whether the result is significant, i.e. whether there is a significant difference between the two samples.

Wilcoxon signed rank test

The **Wilcoxon signed rank test** is another non-parametric test often used to compare **paired** values in two samples. For example, this test can be used to decide whether a medical treatment has had a significant effect. In this case measurements or readings are taken from the patients before and after treatment. These give paired values and the difference between the values in each pair is found.

If the treatment has had no effect then some of these differences will be positive and some will be negative and fairly evenly mixed. If the treatment has had an effect then the readings before treatment should be consistently larger or consistently smaller than those after, giving differences that are consistently positive or consistently negative.

Note that when the treatment has such a large effect that all the readings before are larger than those after (or vice versa), the value of T is 0. Usually the results are not so clear-cut and the value of T must be looked up in a table to determine whether or not the test statistic is significant.

The test statistic, T, is found as follows:

- The differences are ranked in order of size, ignoring their signs.
- The ***ranks*** associated with the positive differences are added to give a total P.
- The ***ranks*** associated with the negative differences are added to give a total Q.
- The test statistic T is the smaller of P and Q.

Chi-squared test

The **χ^2 test** is used to decide whether an observed set of frequencies differs significantly from what is expected. For example, it can be used to compare the actual frequencies from a sample with what would be expected from a particular distribution.

χ is a Greek letter sometimes written as 'chi'. It is pronounced 'keye'.

In **Activity 3.3G** you compared the times taken by a sample of students to travel to college with the distribution of times that would be expected from a sample from a normal distribution with the same mean and standard deviation. The χ^2 significance test can take this one step further by carrying out a test on the differences.

The test statistic is $\chi^2 = \Sigma \frac{(O_i - E_i)^2}{E_i}$ where O_i and E_i represent the observed and expected frequencies of the groups.

The value of the test statistic is compared with a value found from χ^2 tables in order to decide whether or not the theoretical model is a good enough fit for the distribution.

This test is sometimes called a **goodness of fit** test.

The χ^2 test can also be used to test whether two sets of attributes are independent, or whether there is an association between them.

For example, a χ^2 test could be used to test if there is a relationship between the sex of candidates and their ability to pass a driving test. Suppose the results from a group of 60 learners when they take the test are as follows:

Observed	Pass	Fail	Total
Male	13	12	25
Female	20	15	35
Total	33	27	60

Tables of this sort are called **contingency tables.**

The null hypothesis is that gender and ability to pass the test are independent. This implies that the proportion passing the test should be the same for males as for females.

Since altogether 33 out of 60 passed the test,

the ***expected*** number of males to pass $= \frac{33}{60} \times 25 = \frac{33 \times 25}{60} = 13.75$

The table below shows all the expected frequencies.

Expected	Pass	Fail	Total
Male	(33 × 25)/60 = 13.75	(27 × 25)/60 = 11.25	25
Female	(33 × 35)/60 = 19.25	(27 × 35)/60 = 15.75	35
Total	33	27	60

Note that for each cell in the table:

expected frequency $= \dfrac{\textbf{column total} \times \textbf{row total}}{\textbf{grand total}}$

The test is then completed by calculating $\chi^2 = \Sigma \frac{(O_i - E_i)^2}{E_i}$ and comparing the value found with a value found from χ^2 tables.

Using Technology

Throughout this course you will find it useful to use technology to help you not only find the solutions to problems but also to investigate situations and to learn new mathematics.

When working with statistical data, at times, you will find it particularly helpful to use spreadsheet software on a computer. On other occasions you may also find graph-plotting software useful. These types of software are useful when you want to print out your work.

You need to have a graphic calculator when you are working through this book – and also when you are preparing work for your coursework portfolio and when you are sitting the written examination papers.

Graphic calculators

Your graphic calculator is a very powerful tool for analysing data and drawing a limited range of statistical graphs. Only if you have a very large amount of data is spreadsheet software very much more useful.

You can enter your data in a table in the calculator and then use the calculating facilities to find a range of statistical measures. For example, using the data sets of body mass and brain mass of a range of mammals, you can explore their correlation by drawing a scatter diagram and calculating the correlation coefficient and the equation of the regression line.

	Body Mass Mass (kilograms)	Brain Mass Mass (grams)
	x	*y*
Baboon	11	180
Chimpanzee	52	440
Cow	465	423
Deer (roe)	15	98
Donkey	187	419
Giraffe	529	680
Goat	28	115
Gorilla	207	406
Horse	521	655
Jaguar	100	157
Kangaroo	35	56
Pig	192	180
Seal (gray)	85	325
Sheep	56	175
Wolf (gray)	36	120

Data source: **Activity 2.2C – Body mass and brain mass of mammals.**

L1	L2	L3 1
11	180	------
52	440	
465	423	
15	98	
187	419	
529	680	
28	115	

L1(1)=11

1 Data entered table

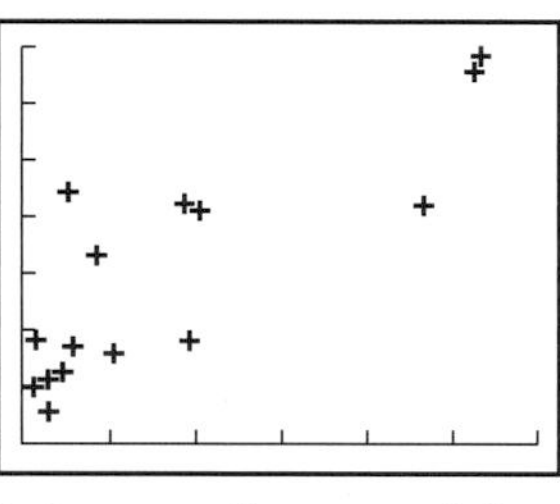

2 Scatter diagram of data

```
EDIT CALC TESTS
1 : 1-Var Stats
2 : 2-Var Stats
3 : Med-Med
4 : LinReg(ax+b)
5 : QuadReg
6 : CubicReg
7↓ QuartReg
```

3 Selecting linear regression calculations

```
LinReg
 y=ax+b
 a=.9051667528
 b=143.2589966
 r2=.7074999101
 r=.8411301386
```

4 Results of linear regression calculations

```
Plot1  Plot2  Plot3
\Y1■.905X+143
\Y2=
\Y3=
\Y4=
\Y5=
\Y6=
\Y7=
```

5 Plotting the regression line

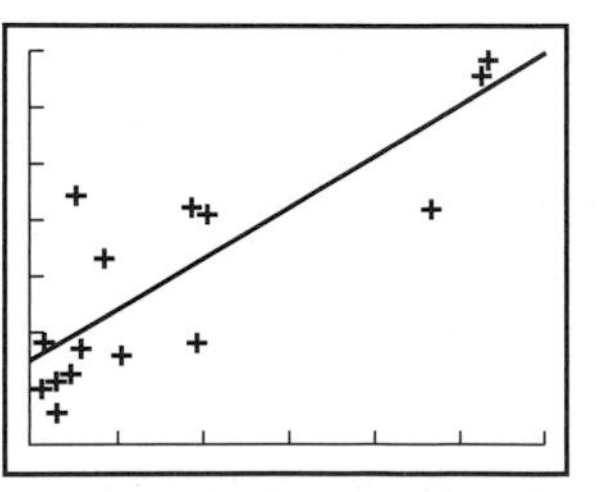

6 Final graph showing data and regression line

Discussion point

Can you explain how the regression line has been plotted?

You can also use the facilities of your graphic calculator to find statistical measures of just one of the data sets or plot a statistical diagram of it showing measures of location and spread.

Discussion points

Which of the two sets of data (body mass or brain mass) is analysed here?
How did you decide?

```
1-Var Stats
 x̄=167.9333333
 Σx=2519
 Σx²=908905
 Sx=186.2948763
 σx=179.9779493
↓n=15
```

Range of statistical measures calculated

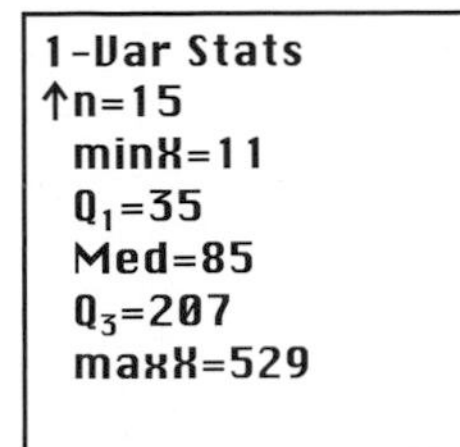

1-Var Stats
↑n=15
minX=11
Q1=35
Med=85
Q3=207
maxX=529

Further statistical measures calculated

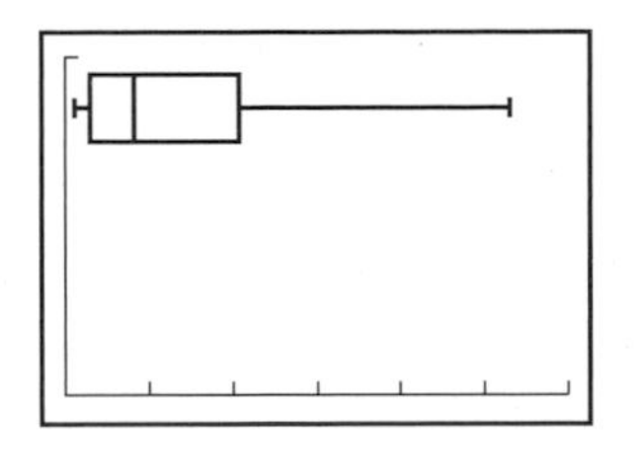

Box plot

Spreadsheets

Spreadsheet software, such as Microsoft Excel, on a computer is not only useful for storing large amounts of numerical data - you can also use the built-in functions to analyse the data and quickly draw statistical diagrams.

By highlighting the data you want to work with, you can, for example, use:

- tools of the spreadsheet to sort the data
- built-in functions to find statistical measures.

Using a spreadsheet to sort data

	A	B	C	D	E
1	Patient ID	Weight at Baseline (kg)	Weight at Week 8 (kg)	Triglyceride level at Baseline (mg/dl)	Triglyceride level at Week 8 (mg/dl)
2	205	69.4	69	143	126
3	223	76.8	73.2	165	96
4	226	81.5	78.9	111	54
5	212	83.4	75	143	102
6	221	83.8	77.9	111	107
7	201	84	82.4	90	131
8	204	84.5	80.4	72	72
9	220	85	80	268	250
10	203	87	81.8	182	152

Data source: **Royal Statistical Society Centre for Statistical Education, Teaching Resource Pack, Statistics in Practice, Case Studies in Medicine.**

Using a spreadsheet to calculate the mean of a data set

1	Region	Alcohol	Tobacco
2	North	£6.47	£4.03
3	Yorkshire	£6.13	£3.76
4	Northeast	£6.19	£3.77
5	East Midlands	£4.89	£3.34
6	West Midlands	£5.63	£3.47
7	East Anglia	£4.52	£2.92
8	Southeast	£5.89	£3.20
9	Soutwest	£4.79	£2.71
10	Wales	£5.27	£3.53
11	Scotland	£6.08	£4.51
12	Northern Ireland	£4.02	£4.56

Data source: **Activity 2.1C – Average weekly household spending on tobacco and alcohol products by region of Great Britain.**

Spreadsheets have a large number of mathematical and statistical functions that you can use to analyse data.

	A	B	C	D	E
1	**Female Students**				
2	130	184	216	75	150
3	215	160	506	240	260
4	565	80	138	205	125
5	321	164	280	415	90
6	132	145	324	634	458
7	346	350	82	136	401

Data source: **Activity 6.1 Use of computers (time in hours per week)**

Discussion point

Do the spreadsheet functions for measures such as median and quartiles give the same values that you calculate by hand?

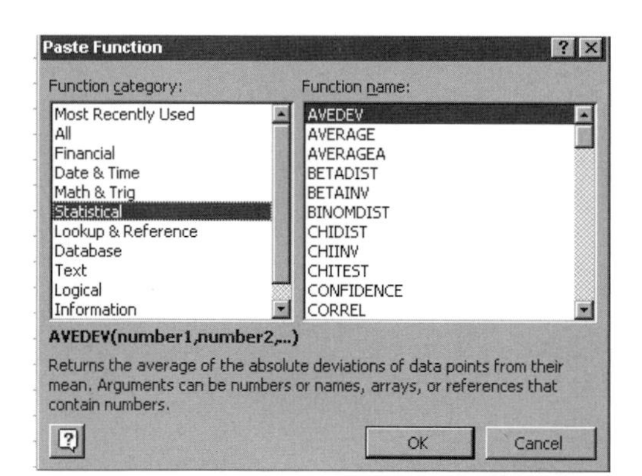

File Edit View Insert Format Tools Data Accounting Wir

Arial 10 B I U

A8 = =MEDIAN(A2:E7)

	A	B	C	D	E
1	**Female Students**				
2	130	184	216	75	150
3	215	160	506	240	260
4	565	80	138	205	125
5	321	164	280	415	90
6	132	145	324	634	458
7	346	350	82	136	401
8	210				

Finding the median of a range of data values.

A small selection of the range of statistical functions that you can paste into a spreadsheet cell.

Of course you can also use a spreadsheet as a very powerful tool to assist you in drawing statistical diagrams. So easy is the process that you have to be careful that you do not draw spurious or erroneous diagrams.

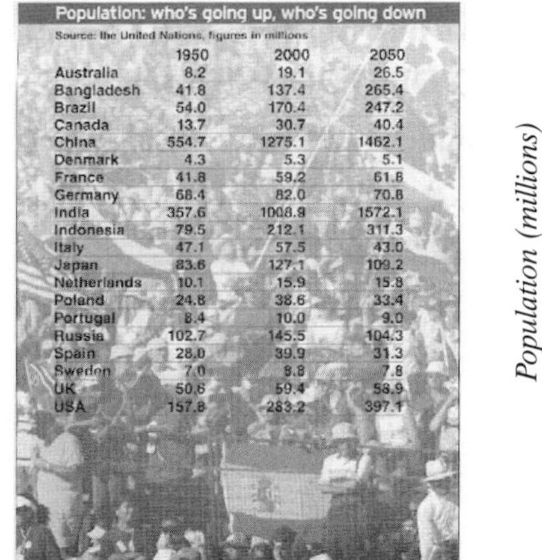

Population: who's going up, who's going down

Source: the United Nations, figures in millions

	1950	2000	2050
Australia	8.2	19.1	26.5
Bangladesh	41.8	137.4	265.4
Brazil	54.0	170.4	247.2
Canada	13.7	30.7	40.4
China	554.7	1275.1	1462.1
Denmark	4.3	5.3	5.1
France	41.8	59.2	61.8
Germany	68.4	82.0	70.8
India	357.6	1008.9	1572.1
Indonesia	79.5	212.1	311.3
Italy	47.1	57.5	43.0
Japan	83.6	127.1	109.2
Netherlands	10.1	15.9	15.8
Poland	24.8	38.6	33.4
Portugal	8.4	10.0	9.0
Russia	102.7	145.5	104.3
Spain	28.0	39.9	31.3
Sweden	7.0	8.8	7.8
UK	50.6	59.4	58.9
USA	157.8	283.2	397.1

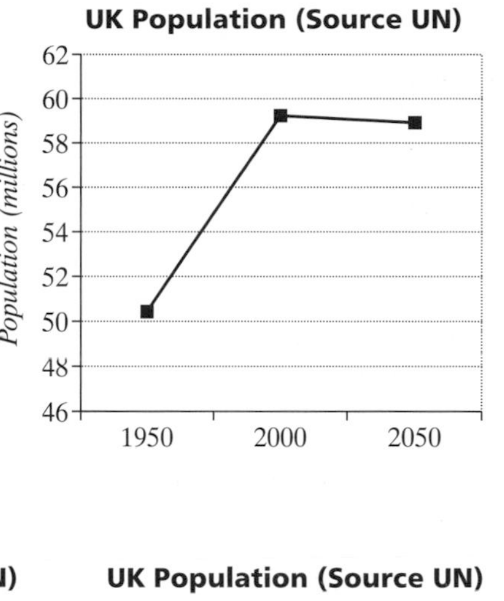

Discussion point

Which of the diagrams is incorrect?
Can you explain your choice?

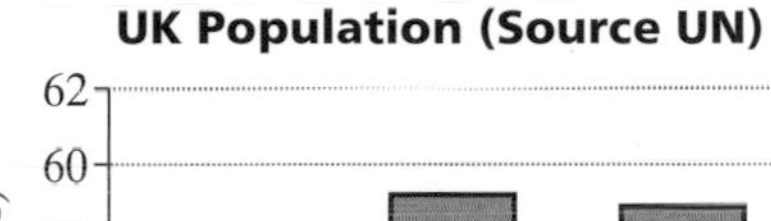

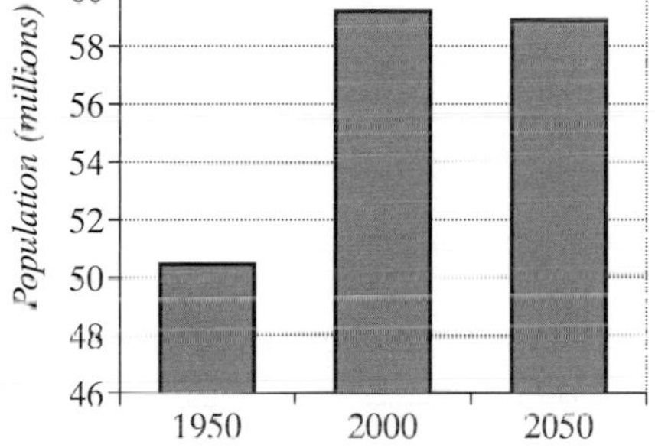

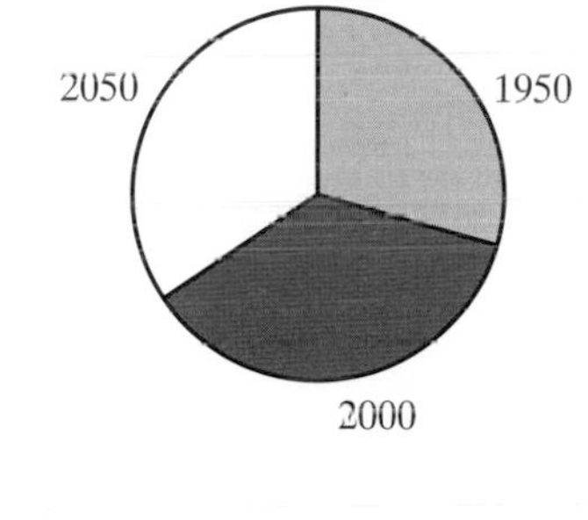

Data source: **Introduction to Chapter 5 – Critical thinking**

Of course it is easy to use spreadsheets to replicate the type of work exploring correlation and regression that was explored using a graphic calculator in this section.

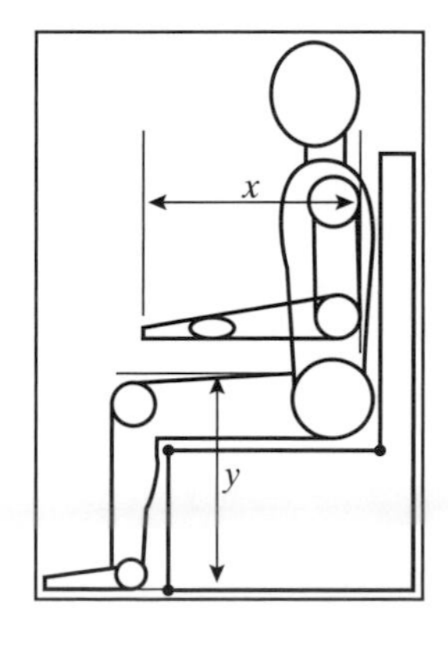

	A	B	C	D
1	Arm and leg data			
2				
3	Boys		Girls	
4	Arm (cm)	Leg (cm)	Arm (cm)	Leg (cm)
5	36	48	39	43
6	37	44	39	44
7	37.5	45	40	48
8	38	53	40	49
9	39	45	41	45
10	39	47	41	48
11	39	47	41	49
12	40	47.5	41	50
13	40	50	42	43
14	41	60	42	46
15	42	53	42	49
16	43	48	42	50
17	43.5	51	43	45
18	44	53	43	49
19	45	53	43.5	45

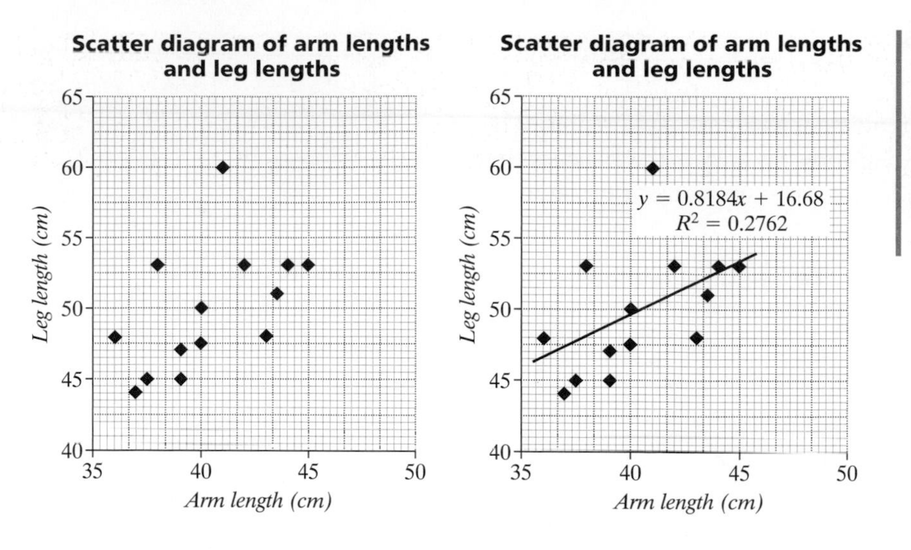

Discussion point

In the second graph a regression (trend) line has been added and its equation is given together with a value of R^2. What is the correlation coefficient in this case? How did you find this?

Graph plotting software

You may wish to use graph plotting software on a computer to plot statistical diagrams and calculate some statistical measures. Software such as *Autograph* is very easy to use to produce such graphs.

Region	Alcohol	Tobacco
North	£6.47	£4.03
Yorkshire	£6.13	£3.76
Northeast	£6.19	£3.77
East Midlands	£4.89	£3.34
West Midlands	£5.63	£3.47
East Anglia	£4.52	£2.92
Southeast	£5.89	£3.20
Southwest	£4.79	£2.71
Wales	£5.27	£3.53
Scotland	£6.08	£4.51
Northern Ireland	£4.02	£4.56

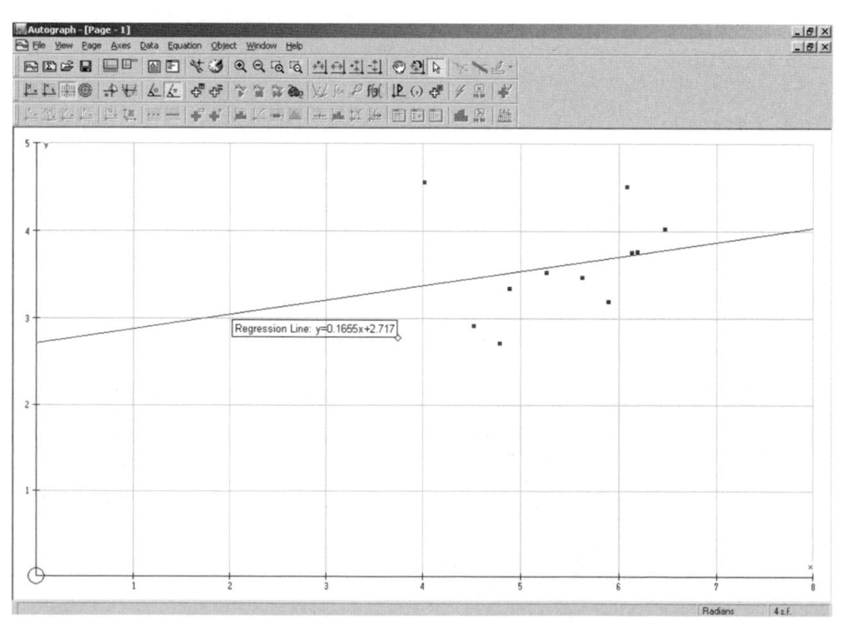

Data source: **Activity 2.1C – Average weekly household spending on tobacco and alcohol products by region of Great Britain.**

Answers

1 Exploring and describing data

1.1 Collecting data

1.1C

1

	<16	16–59	60+
Male	114	290	90
Female	112	293	101

2 a 20–99: 3287; 100–499: 3553; 500–999: 1644

1.1D

1 Choices do not cover all possibilities or allow for them, e.g. red hair, dyed green hair. Include a box for 'other' with room for details.
2 Is 1 (or 5) very greasy, or not at all greasy? Include description for 1 and 5, so the scale is obvious.
3 Available answers not the choices needed in question! Allow, e.g. 'shampoo', 'conditioner', 'both', 'neither'.
4 'Less than once' … is this per week? Per month? Per year? Maybe include 'more than once a day'.
5 'Up to £2' and '£1–£3' overlap. Change the first option to 'Up to 99p. Other groups also overlap at £3'.
6 Will people understand what each ingredient is supposed to do to hair? What is meant by 'more effective'? Will that term prompt people to select all options?! Omit question.

1.2 Measures of location and spread

1.2A

1a, 2a, 3a

Team	Mode	Median	Mean
Argentina	1	1	0.67
Belgium	N/A	1.5	1.50
Brazil	2	2	2.57
Cameroon	1	1	0.67
China	0	0	0.00
Costa Rica	2	2	1.67
Croatia	0	0	0.67
Denmark	2	1.5	1.25
Ecuador	1	1	0.67
England	1	1	1.20
France	0	0	0.00
Germany	1	1	2.00
Ireland	1	1.5	1.75
Italy	1	1	1.25
Japan	2	1.5	1.25
Korea	2	2	1.86
Mexico	1	1	1.00
Nigeria	0	0	0.33
Paraguay	N/A	1.5	1.50
Poland	0	0	1.00
Portugal	N/A	2	2.00
Russia	2	2	1.33
Saudi Arabia	0	0	0.00
Senegal	1	1	1.40
Slovenia	1	1	0.67
South Africa	2	2	0.83
Spain	3	3	3.00
Sweden	1	1	1.25
Tunisia	0	0	0.33
Turkey	1	1	1.43
United States	1	1	1.40
Uruguay	N/A	1	1.33

Some teams scored a different number in each game.

1 b i Spain
ii Can't say from the table alone, as you have no information on quality of opponents faced. Is 'best at scoring goals' purely the number scored or the best shot/goal ratio?
c A mean or median value could be decimal unlike the modal value, so the mode could be said to give the 'typical' score.
2 b i Spain
ii See **1b ii**
c Not all teams have a modal value, but will have means and medians. However, median values are less affected by the odd unusual outcomes (e.g. high scoring matches against weak teams).
3 b No. For the small number of games here, 8 will raise the mean score disproportionately.
c i No, cannot have a decimal number of goals
ii Two games had extra time, one being decided on penalties
d i Spain
ii See **1b ii**
e Any with suitable reason.

1.2B

1 Ferrari
2 a

	Mean*	Median
Ferrari	143	133
McLaren	102	102
Williams	83	80

*Given to nearest points

b Either with suitable reason (e.g. mean includes all years).
3 Ferrari (range of scores is similar to McLaren, but leaving out the highest and lowest, Ferrari's scores lie in a smaller range of about 70 points).
4 a

	Range
Ferrari	151
McLaren	107
Williams	140

b i McLaren
ii Depends on student's answer to 2b.
5 a

	IQR
Ferrari	77
McLaren	89
Williams	87

b i Ferrari
ii McLaren
c Ferrari
6 b,c Mean difference will be zero
d iii 2206 Yes, see **6c**

7 a

	SD*
Ferrari	47.0
McLaren	40.7
Williams	48.8

*to 3 s.f.

c i Williams **ii** Ferrari

8 a

Team	Range	IQR	SD
Argentina	1	1	0.47
Belgium	3	2.5	1.12
Brazil	4	2	1.29
Cameroon	1	1	0.47
China	0	0	0.00
Costa Rica	1	1	0.47
Croatia	2	2	0.94
Denmark	2	1.75	0.83
Ecuador	1	1	0.47
England	3	1.5	0.98
France	0	0	0.00
Germany	8	1	2.51
Ireland	2	1.75	0.83
Italy	1	0.75	0.43
Japan	2	1.75	0.83
Korea	2	1	1.46
Mexico	2	1.5	0.71
Nigeria	1	1	0.47
Paraguay	3	2.5	1.12
Poland	3	3	1.41
Portugal	4	4	1.63
Russia	2	2	0.94
Saudi Arabia	0	0	0.00
Senegal	2	2	1.02
Slovenia	1	1	0.47
South Africa	1	1	0.47
Spain	0	0	0.00
Sweden	1	0.75	0.43
Tunisia	1	1	0.47
Turkey	3	2	1.05
United States	3	2	1.02
Uruguay	3	3	1.25

b None, because of inclusion of extra time.

1.2C

Birmingham: Mean = 16.2; SD = 23.4 (min)

1 a **i** Manchester: Mean = 23.5; SD = 38.6 (min)
ii Newcastle: Mean = 18.9; SD = 34.6 (min)

c **i** Manchester **ii** Birmingham

d **i** Manchester **ii** Birmingham

2 'Early–15' for all 3 airports.

3 Not with any certainty. Impossible to tell where in the ranges the maximum and minimum times fell.

4 a Early flights are counted as zero delay ⇒ more flights counted at zero (not uniform distribution through group)

b **i** 9067.5; Early–15 **ii** 2247; Early–15

c **i** 10.9 minutes **ii** 9.8 minutes

1.2D

1 a £26 122.02 (to nearest pence)

b £18 930 **c** £24 975.50

d £5289.67 **e** £6797.25

2 a

Salary (£000s)	Frequency
20	16
25	23
30	11
35	8
40	2
Total	60

b £25 000 (to nearest £5000)

c **i** £26416.67 **ii** £5484.19

iii £25652.17 **iv** £8302.56

3 a

Salary (£000s)	Frequency
15–	12
20–	20
25–	16
30–	9
35–	3
Total	60

b £20 000–£25 000

c **i** £25 083.33 **ii** £5589.55

iii £24 625.00 **iv** £8484.38

4 b Yes – an average cannot be calculated accurately without the original data.

1.3 Statistical charts and graphs

1.3A

1

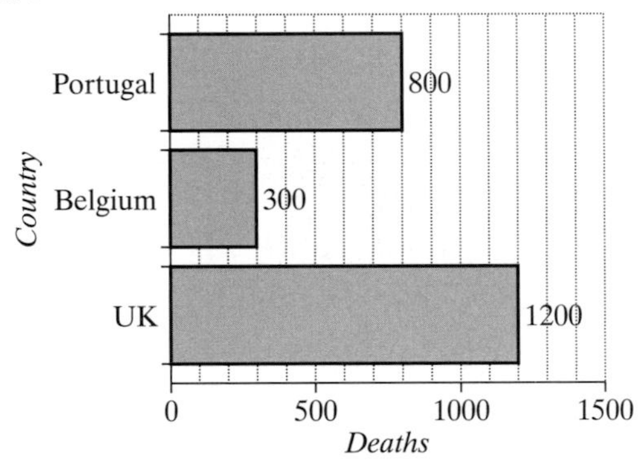

3 a Accept estimates that are similar to these:

Age	Population (millions)	KSI Pedestrian	KSI Cars	KSI Other
50–59	7.70	680	2260	1160
60–69	5.90	730	1530	470
70–79	4.20	880	1144	290
80+	2.60	850	660	160
Total	20.40	3140	5590	2080

1.3B

2 a

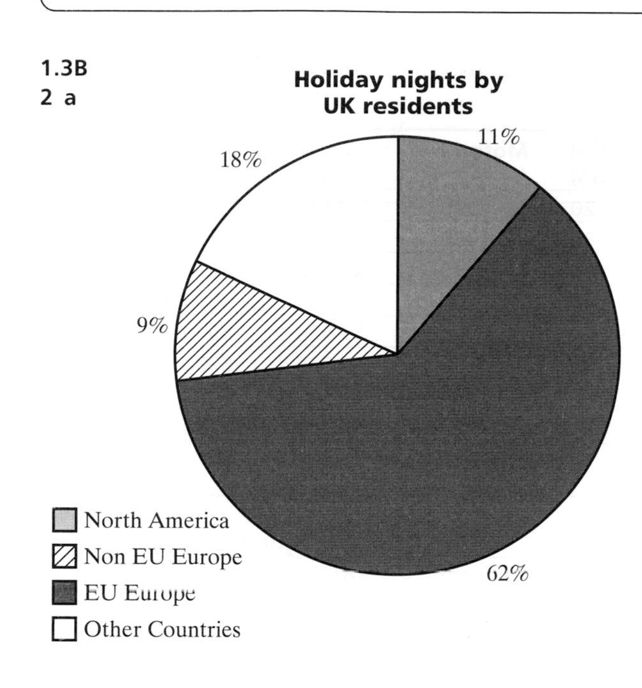

Business nights by UK residents

21%
17%
9%
53%

North America
Non EU Europe
EU Europe
Other Countries

3 a

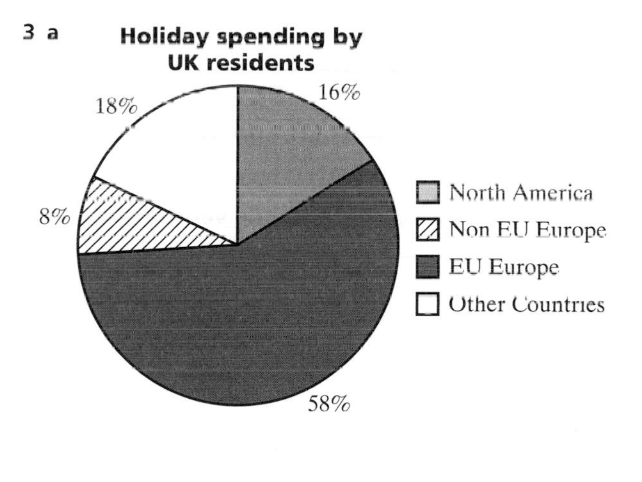

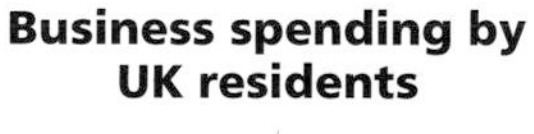

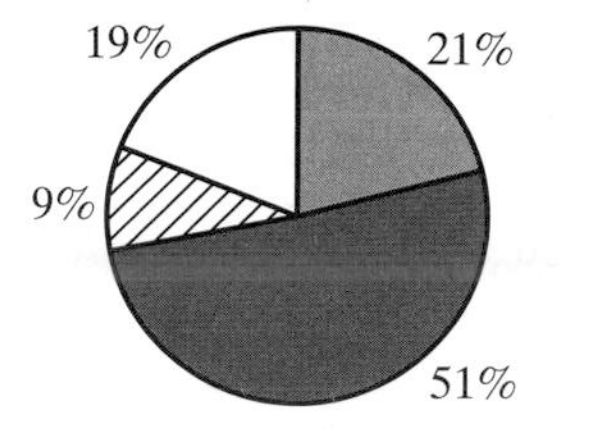

1.3C

1a, b

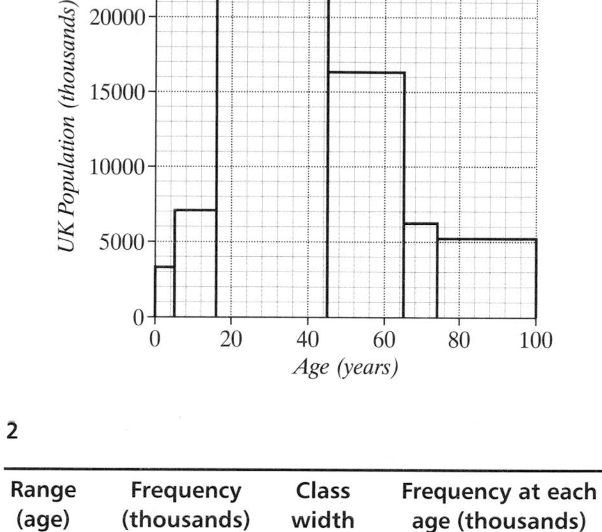

2

Range (age)	Frequency (thousands)	Class width	Frequency at each age (thousands)
0–4	3299	5	660
5–15	7078	11	643
16–44	21 775	29	751
45–64	16 316	20	816
65–74	6239	10	624
75+	5219	25	209

(i.e. 660 000 aged 0, 1, 2, 3 and 4;
643 000 aged 5, 6, 7 … etc.)

3

Age (years)	Frequency (thousands)	Lower boundary	Upper boundary	Class width	Frequency density (thousands per year)
0–4	3299	0	5	5	660
5–15	7078	5	16	11	643
16–44	21 775	16	45	29	751
45–64	16 316	45	65	20	816
65–74	6239	65	75	10	624
75+	5219	75	100	25	209

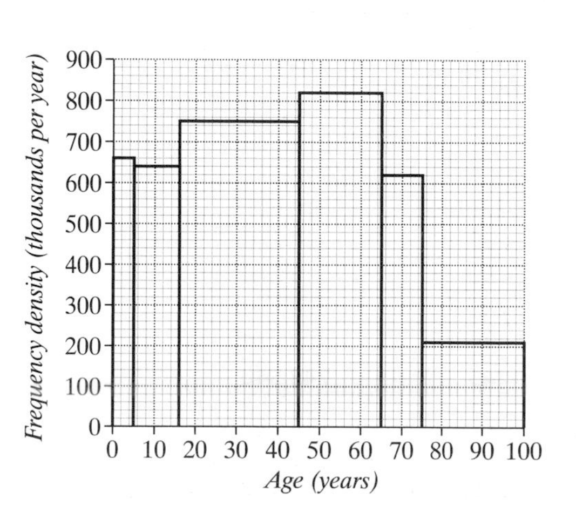

7 a

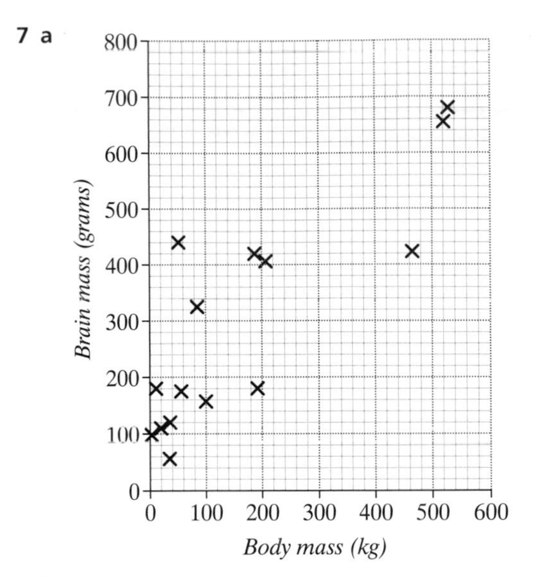

b 0.841

c As body mass increases, brain mass increases

d **i** 168 kg

ii 295 g

e **i**

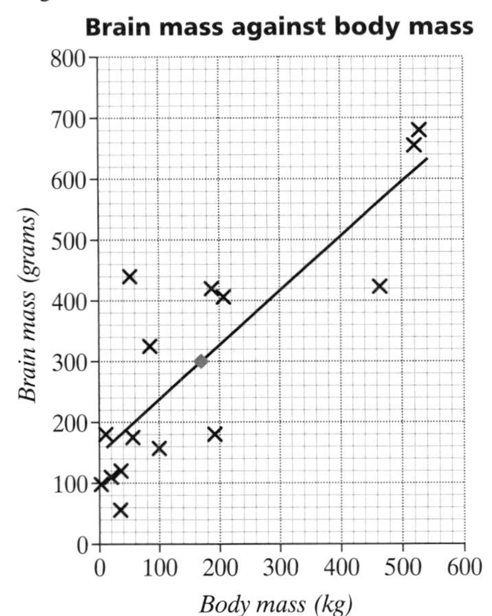

ii 211 g

iii 143 g – makes no sense as you can't have negative weight!

2.3 Regression

2.3A

2 47% approx.

3 Accept lines with similar gradients and intercepts to $g = -1.2s + 65$ where $g = -1.2s + 65$, where g is the percentage of pupils gaining 5 or more A*–C grade GCSEs and s is the percentage of pupils receiving free school meals.

4 Same – allowing for reading errors off the graph.

2.3B

1 $a = 58.3$; $b = 0.002\,45$

$y = -1.18 + 65.1$ (3 s.f.)

a,b

Country	GDP per capita in dollars (X)	Medal points in 1986 Olympics (Y)	XY	X squared
United States	33 900	221	7 491 900	1 149 210 000
Russia	4200	136	571 200	17 640 000
Germany	22 700	123	27 792 100	515 290 000
China	3800	104	395 200	14 440 000
France	23 300	74	1 724 200	542 890 000
Italy	24 100	71	1 711 100	580 810 000
Australia	22 200	68	1 509 600	492 840 000
South Korea	13 300	56	744 800	176 890 000
Cuba	1700	51	86 700	2 890 000
Ukraine	2200	43	94 600	4 840 000
Totals	151 400	947	17 121 400	3 497 740 000

c $a = 0.002445$, $b = 58.3$ to 3 s.f.

d

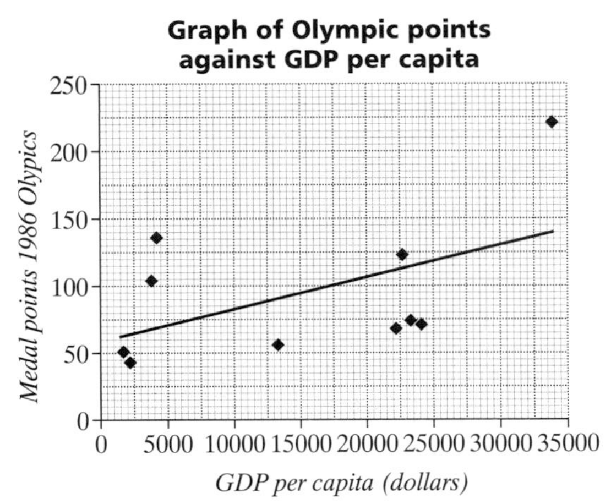

e regression line fits quite well (reference to correlation coefficient, $r = 0.51$)

f the number of medal points a country with a GDP per capita of 0 would have got at the 1986 Olympics when using the regression line as the model.

g the additional number of medal points a country gets for each additional dollar's GDP/capita.

3 a Mean temperature: 74°F (73.8°F);

Mean humidity: 38.4%

b **i** $y = -1.878x + 177.07$

c **i** 38.1%

ii 17.4%

iii First estimate, because it is within the range of data on which the regression line is drawn.

d Because actual data at temperatures above and below where the line would suggest, have humidities below 35% (A curve would fit the data better).

4 b ii The greater the median annual earning, the lower the unemployment rate.

c **i** Mean earnings: \$41 412.50; Mean rate: 2.875%

ii $y = -0.000\,084\,7x + 6.38$

c iii

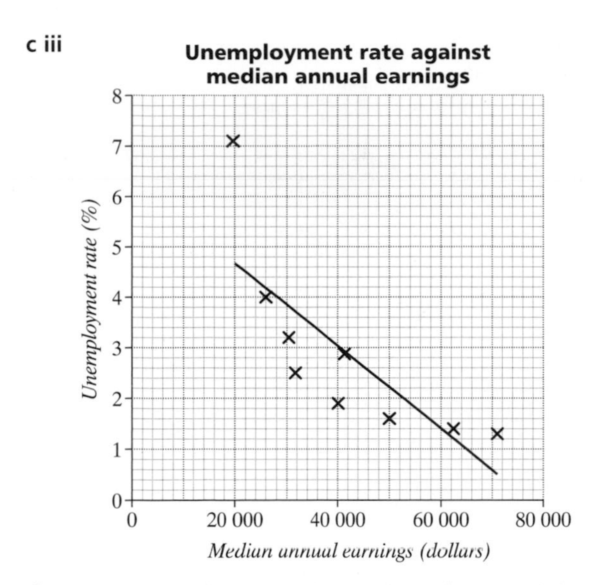

d Data obviously fits a curve better than a linear model.

2.3C

2 a 50.2 cm **b** 57.6 cm

3 a Equation $y = 0.744x + 19.0$ (3 s.f.)

b

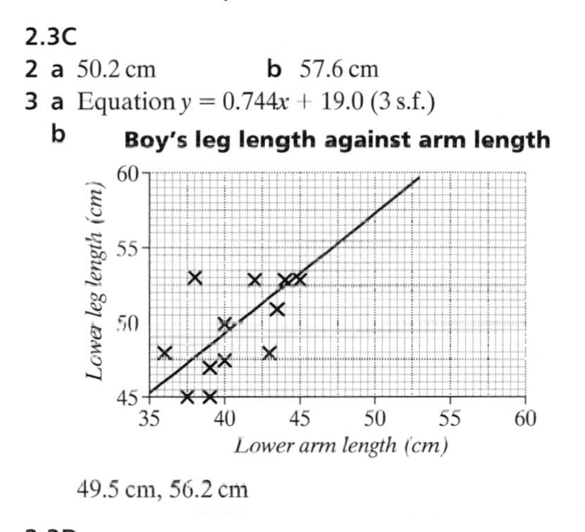

49.5 cm, 56.2 cm

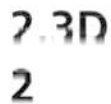

2.3D

2

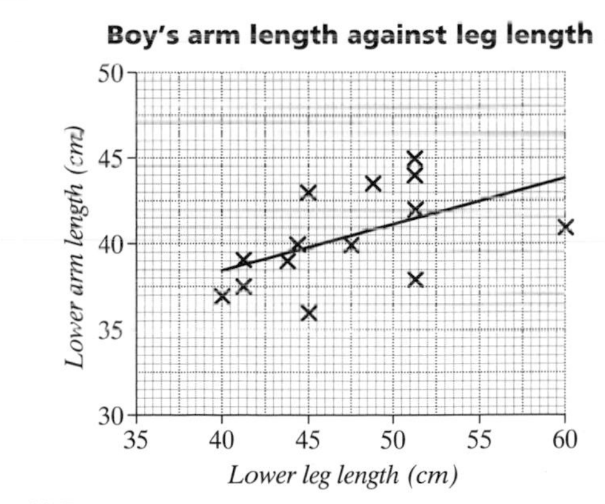

3 a 40.7 cm

b Leg lengths from between 40–52 cm, i.e. not much beyond the range of the data.

2.5 Preparing for assessment

Practice exam questions

1 a **i** 64.8
ii 84.1
iii 0.750
iv $y = 0.529x + 49.8$

b

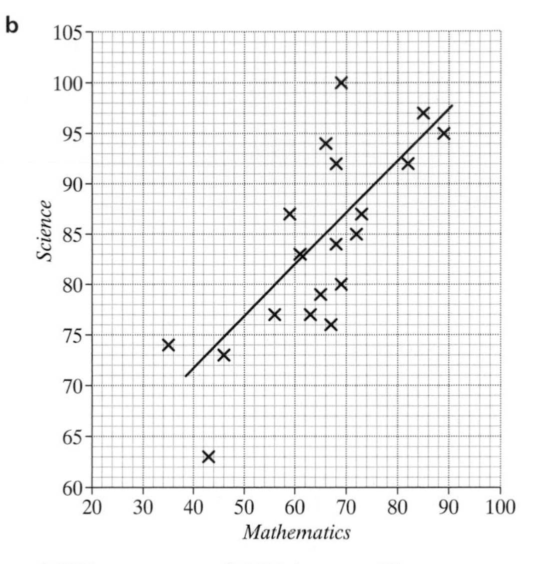

c **i** 60% **ii** 76% (nearest %)

d Maths/science because the correlation is stronger (i.e. c(ii)).

e There appears to be a stronger correlation between maths and science than between maths and English.

2 a **i** 85.1 g **ii** 93.8 g
iii $r = 0.994$ **iv** $y = 1.38x - 23.8$

b 252 g

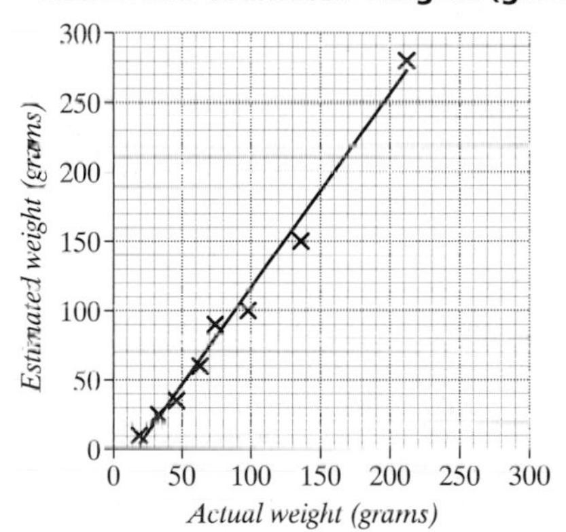

c The gradient gives the factor at which the estimates are overestimating each additional (actual) gram.

3 The normal distribution

3.1 Samples from normal and other distributions

3.1A

1 a i

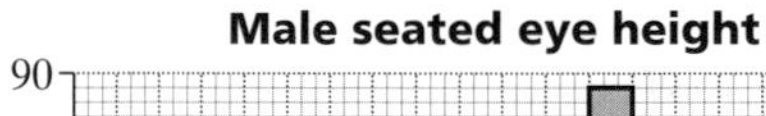

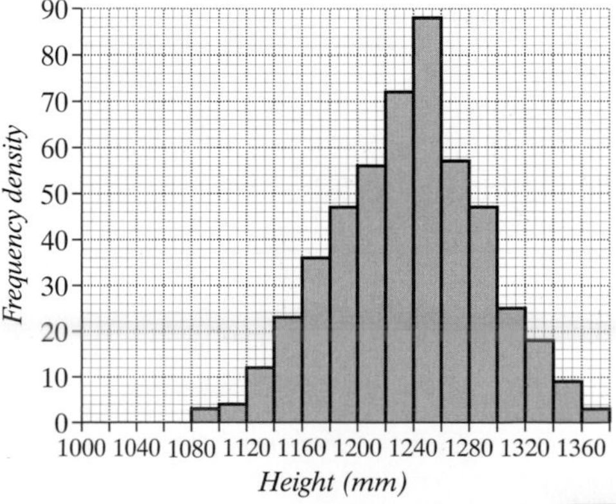

ii

iii

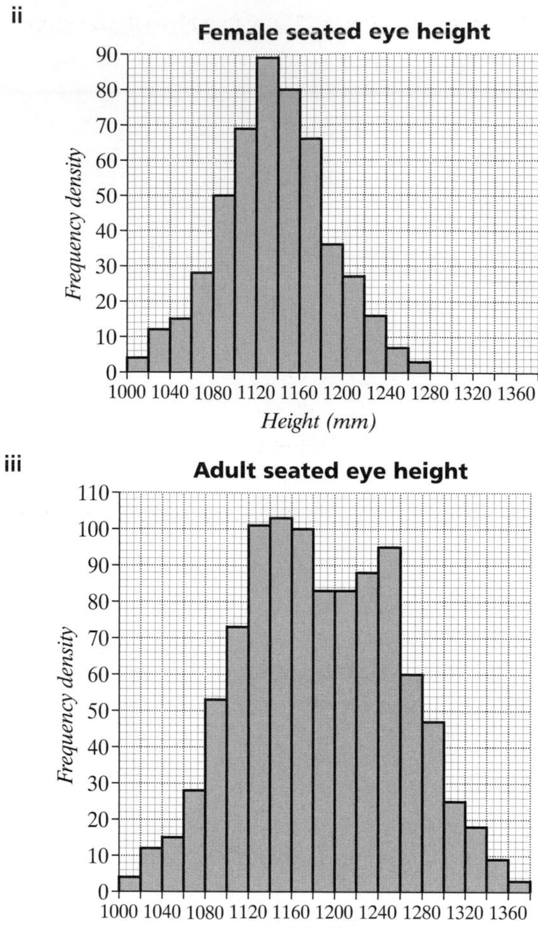

b Male and female histograms are similar in shape, with the majority of data within the central 90 cm. The adult histogram has 2 distinct peaks.

2 a i

ii

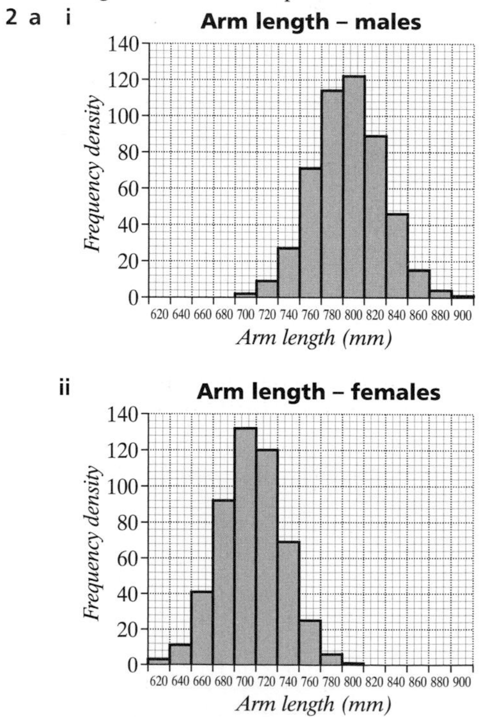

iii

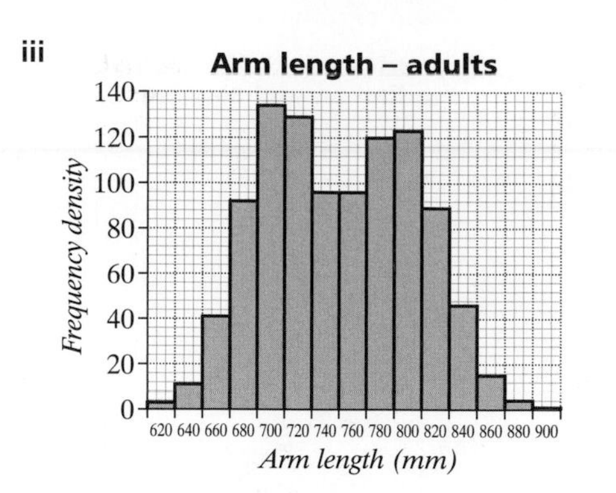

b As in **1b**

3 a Mean = 1236; SD = 53.2 mm
b Mean = 1137; SD = 49.3 mm
c Mean = 1186; SD = 71.4 mm
d Mean = 805; SD = 32.2 mm
e Mean = 716; SD = 29.7 mm
f Mean = 760; SD = 54.2 mm

4 a modal class: 1240–1260 mm; median 1239 mm
b modal class: 1120–1140 mm; median 1136 mm
c modal class: 1140–1160 mm; median 1183 mm
d modal class: 800–820 mm; median 804 mm
e modal class: 700–720 mm; median 716 mm
f modal class: 700–720 mm; median 759 mm

3.1B

1 a 1326 mm
b **i** 1052 mm
ii 1326 mm
iii 274 mm to nearest mm.

2 a **i** 665 mm
ii 858 mm
b **i** 678 mm
ii 847 mm
c 667 mm – 858 mm to nearest millimetre.

3.1E

1 a i

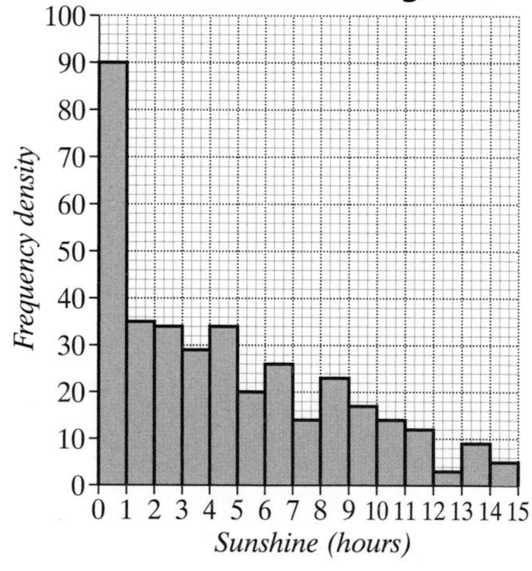

ii (Highly) positively skewed

b i

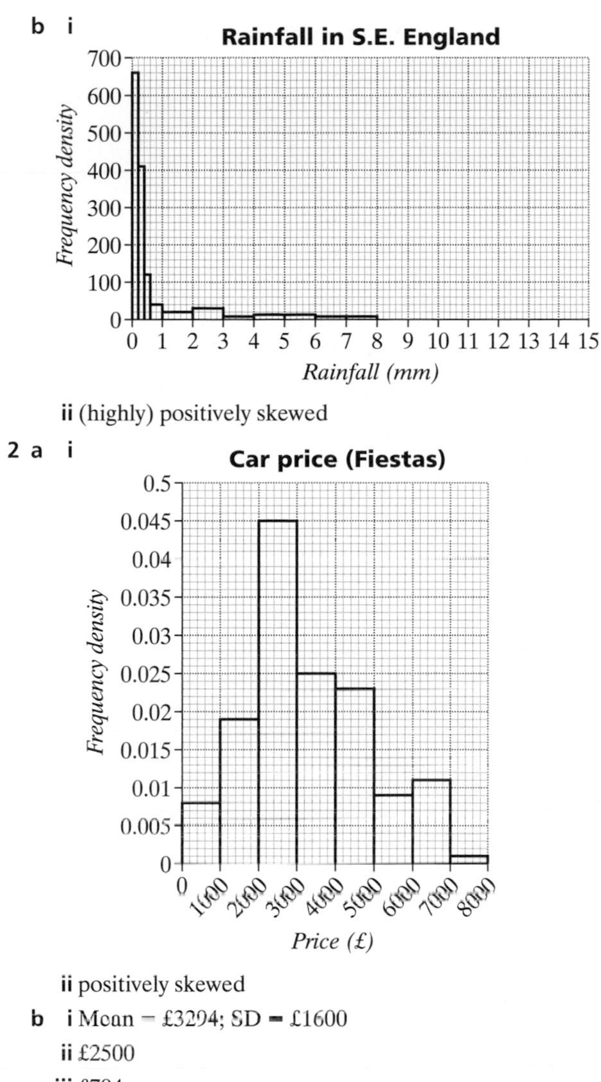

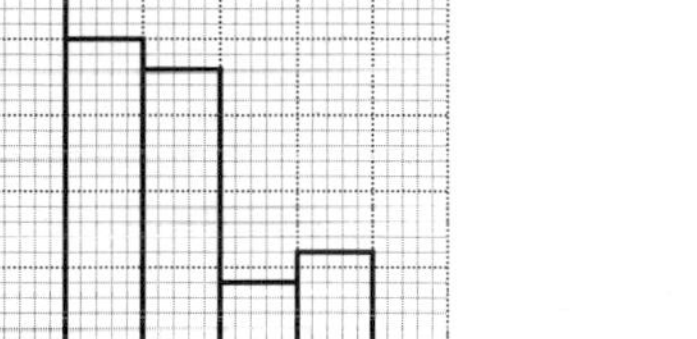

ii (highly) positively skewed

2 a i

ii positively skewed

b i Mean = £3294; SD = £1600

ii £2500

iii £794

c i 102

ii 97 cars

3 a

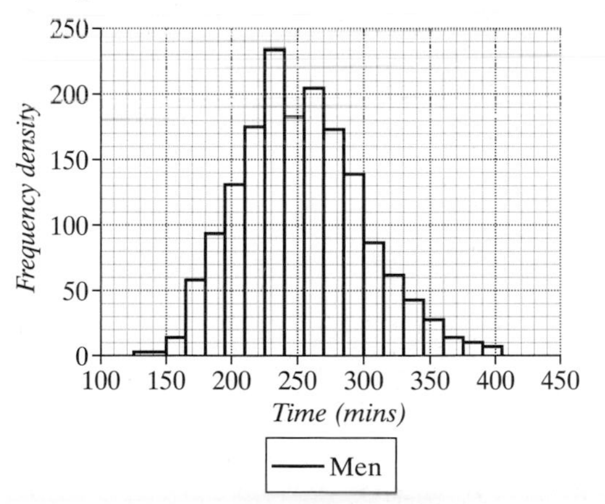

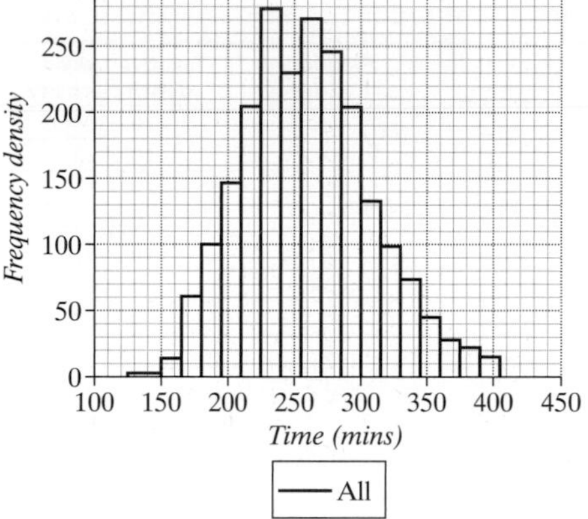

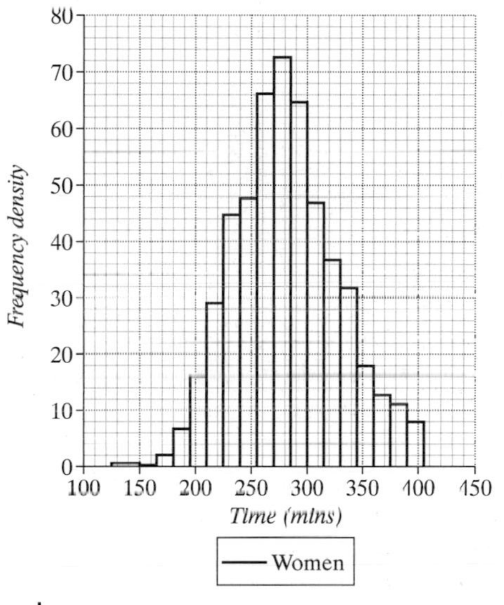

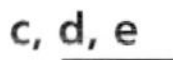

c, d, e

	Men	Women	All
Mean	4:13	4:43	4:20
Median	4:10	4:40	4:17
Difference	3 mins	3 mins	3 mins

f 12 900 (3 sf)

g i 4106 (3 sf)

ii Do not know the distribution of the 11 women running between 2:05 and 2:29.

4 a $0 \leq t < 30$ **b** 43 s

c i

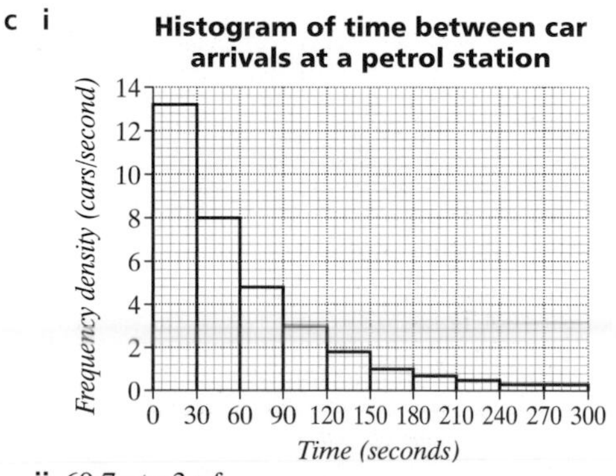

ii 60.7 s to 3 s.f.

3.2 The Standard Normal Distribution

3.2A

1 a

Sitting eye height	Female (mm)	Frequency frequency	Relative freqency density density
1000–	4	0.2	0.0004
1020–	12	0.6	0.0012
1040–	15	0.75	0.0015
1060–	28	1.4	0.0028
1080–	50	2.5	0.005
1100–	69	3.45	0.0069
1120–	89	4.45	0.0089
1140–	80	4	0.008
1160–	64	3.2	0.0064
1180–	36	1.8	0.0036
1200–	27	1.35	0.0027
1220–	16	0.8	0.0016
1240–	7	0.35	0.0007
1260–	3	0.15	0.0003
Total	500		

b

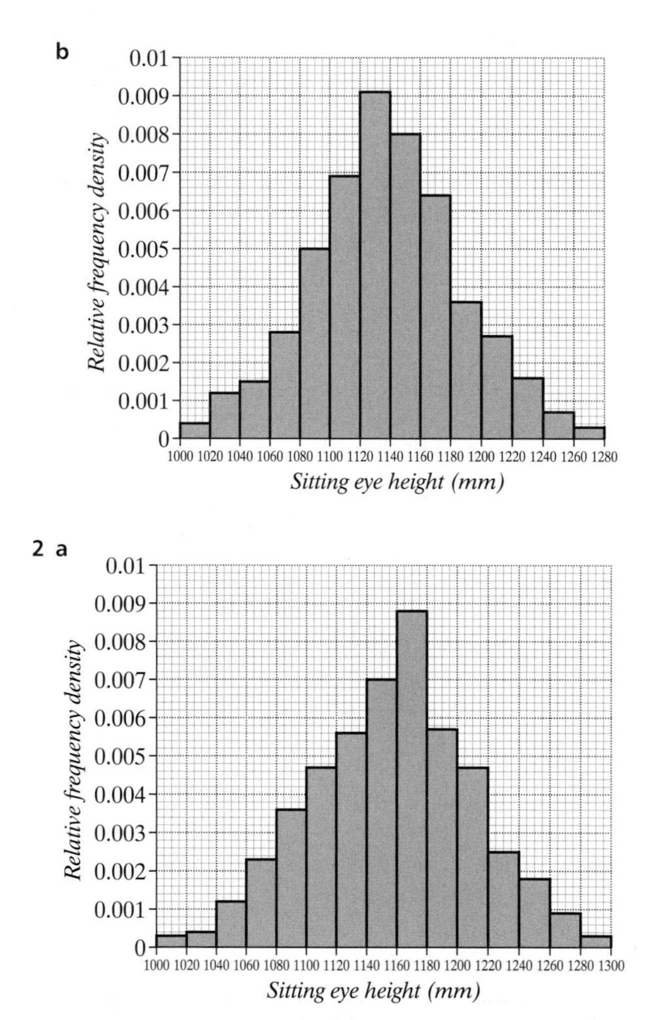

b Similar shapes and relative frequency densities. Different eye heights.

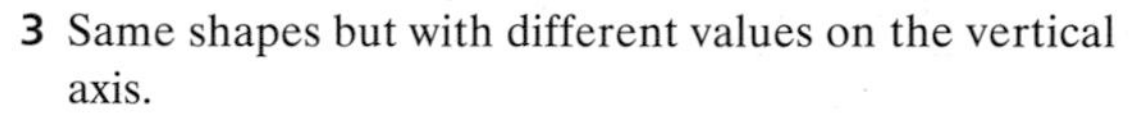

3 Same shapes but with different values on the vertical axis.

3.2B

2 a $\mu = 40, \quad \sigma = 10$

b $\mu = 20, \quad \sigma = 5$

c $\mu = 500, \sigma = 50$

d $\mu = 0, \quad \sigma = 10$

3 a

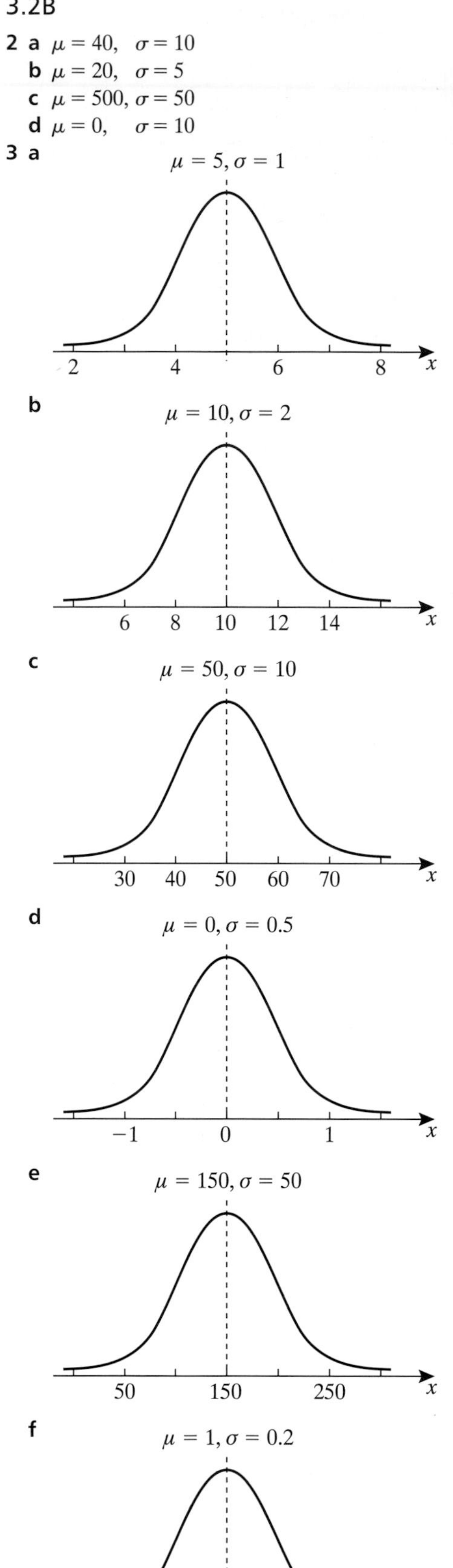

4 a

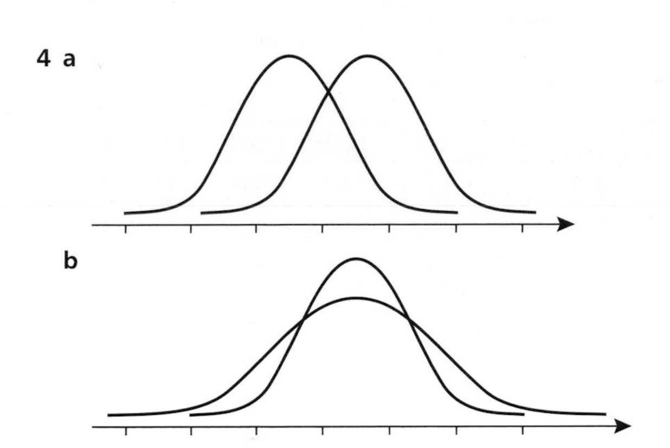

b

5 a

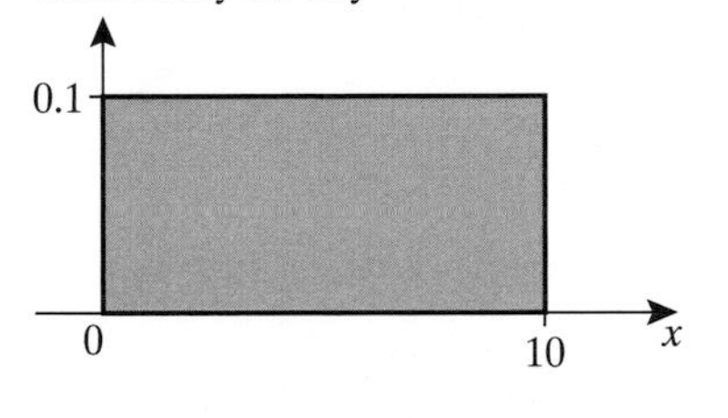

b

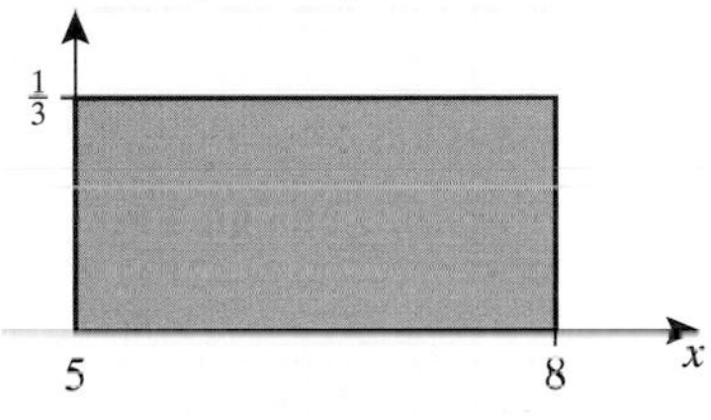

c

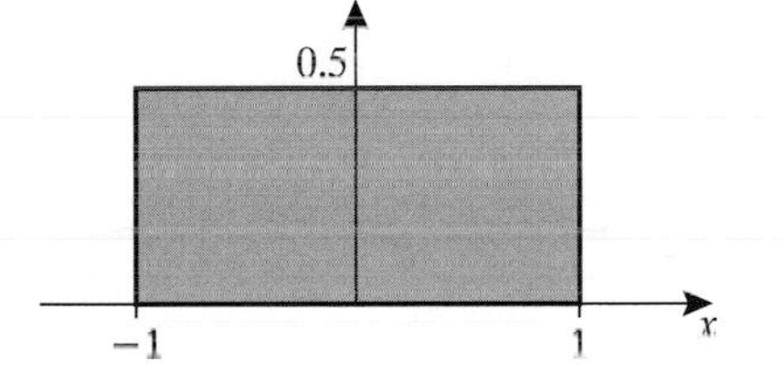

3.2C

1 0.9599
2 0.0808
3 0.8413
4 0.0089
5 0.9484
6 0.0228

3.2D

1 0.4987
2 0.4131
3 0.0586
4 0.0683
5 0.6826
6 0.9544
7 0.9974
8 0.7175
9 0.6127
10 0.4021

3.3 Other normal distributions

3.3A

1 94%
2 Very near 100%
3

	Octopus (<150 cm)		Calgary (>110 cm)	
Age	**Boys**	**Girls**	**Boys**	**Girls**
5	100%	100%	70.7%	63.7%
6	100%	100%	98.8%	97.8%
7	100%	100%	99.9%	99.9%
8	100%	100%	100%	100%
9	99.8%	99.7%	100%	100%
10	93.1%	94.0%	100%	100%
11	70.5%	57.8%	100%	100%
12	36.7%	24.0%	100%	100%
13	10.5%	6%	100%	100%
14	2.6%	1.9%	100%	100%
15	0%	6%	100%	100%

3.3B

1 a 84.1% **b** 6.7% **c** 1.2%

2

Stated length	Probability tape is shorter than stated length
1 hour	15.9%
2 hours	10.6%
3 hours	4.8%
4 hours	4.8%

3 a 0.6% **b** 2.9% **c** 47.1%
4 Triple Jumper: 5.7%; Javelin Thrower: 2.2%; 100 m Sprinter: 55.2%; 1 mile runner: 28.4%
5 A: 7.8%; B: 35.56%; C: 49.9%; Fail: 6.7%
6 **Numbers systolic**

	Mild	Moderate	Severe
Men	319	124	1
Women	265	116	1

Numbers diastolic

	Mild	Moderate	Severe
Men	126	9	1
Women	79	5	0

3.3C

1

Deck	Class	Numbers between 70 and 90 kg	Total
Upper	Business	18	42
	Economy	19	45
Main	Business	4	10
	Economy	61	145

2 a Passengers between 90 kg and 110 kg

Deck	Class	Male	Female	Total
Upper	Business	11	4	15
	Economy	12	4	16
Main	Business	3	1	4
	Economy	40	13	53

b Passengers > 110 kg

Deck	Class	Male	Female	Total
Upper	Business	1	0	1
	Economy	1	0	1
Main	Business	0	0	0
	Economy	4	1	5

3.3D

1 76 mm–80 mm

2

Size	Min. length (mm)	Max. length (mm)
Extra small	170	182
Small	182	187
Medium	187	192
Large	192	198
Extra large	198	209

3

Size	Min. width (mm)	Max. width (mm)
Extra small	80	87
Small	87	90
Medium	90	93
Large	93	97
Extra large	97	103

Assuming that shortest and narrowest hands occur in the same group.

3.3E

1056 mm, 1323 mm

3.3F

1 a 51 min 39 sec
b 53 min 35 sec
c 57 min 7 sec

2 a Boys

Age	UQ	LQ	Percentile 10th	90th	3rd	97th
4	19.4	16.6	15.3	20.7	14.1	21.9
8	29.9	24.1	21.5	32.5	18.9	35.1
12	45.8	35.2	30.4	50.6	25.6	55.4
16	67.7	55.3	49.6	73.4	44.0	79.0

Girls

Age	UQ	LQ	Percentile 10th	90th	3rd	97th
4	19.1	15.9	14.4	20.6	13.0	22.0
8	30.1	23.9	21.0	33.0	18.2	35.8
12	47.6	36.4	31.4	52.6	26.4	57.6
16	60.5	50.5	45.9	65.1	41.4	69.6

3 Standard: 1100 hrs; Low Energy: 10 900 hrs; Fluorescent: 8900 hrs

4 a $\dfrac{454 - \mu}{\sigma} = -0.93$ and $\dfrac{460 - \mu}{\sigma} = 1.46$

4 b $\mu = 456$ g, $\sigma = 2.51$ g (to 3 s.f.)

4c 6%

3.3G

1

Histogram of journey times for students

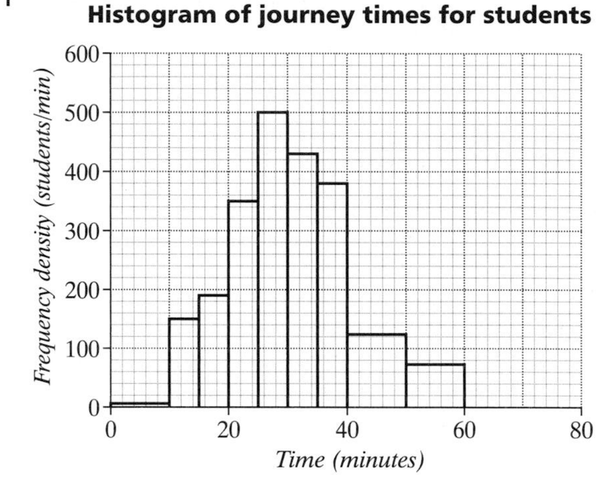

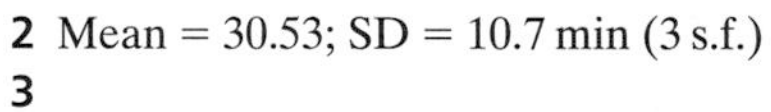

2 Mean = 30.53; SD = 10.7 min (3 s.f.)

3

SDs from mean	Lower	Upper	Proportion	Normal Distribution
1	19.83	41.23	68.9%	68.3%
2	9.13	51.93	95.3%	95.4%
3	−1.57	62.63	99.9%	99.7%

4

Time	Expected
0–9	298
10–14	510
15–19	1014
20–24	1627
25–29	2105
30–34	2199
35–39	1854
40–49	1955
50–59	417
60+	39

5 i

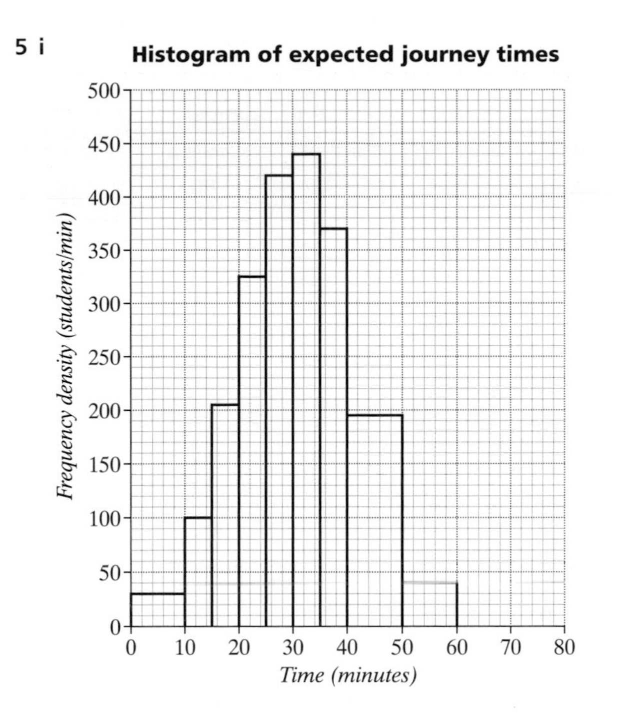

3.5 Preparing for assessment

Practice exam questions

1 a 9-year-old boys have a larger mean head circumference and greater variability/spread than 9-year-old girls.
b i, ii 53 cm
c i 57%
ii Sample children from other racial backgrounds, and include both urban and rural children.

2 a i, ii

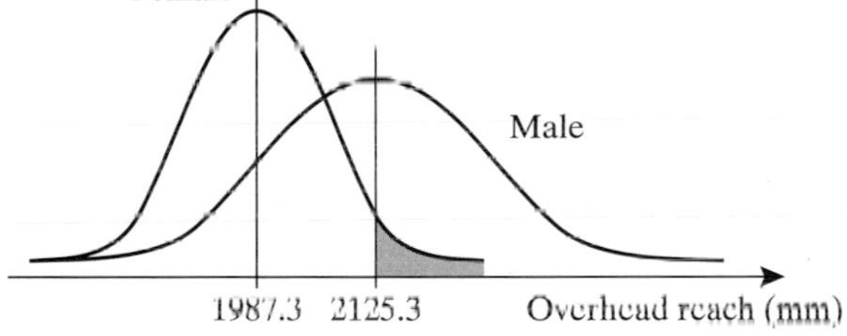

iii 4%
b i 2260 mm **ii** 1860 mm

3 a i peak at centre, symmetry
ii 95% between $\mu + 2\sigma$, 50% above μ, etc.
b very large 10%
large 56%
medium 32%
small 2%
(to nearest %)
c 36 eggs

4 Interpreting tables and diagrams

4.1 Tables of data

4.1A

1 a i Estonia **ii** Denmark
b i France **ii** Russia
c Estonia
d Denmark

2 a i $\frac{40\,383 - 24\,915}{24\,915} \times 100\%$
ii $\frac{40\,383 - 37\,322}{37\,322} \times 100\%$
b i $\frac{19\,956 - 21\,272}{21\,272} \times 100\%$
ii $\frac{19\,956 - 21\,788}{21\,788} \times 100\%$

3 a Czech Republic, Estonia, Hungary, Poland, Greece, Ireland, Netherlands, Russia, Slovenia
b Denmark, Sweden

4 No. The decrease in 1998 is 8% of 21 788 = 1743, but an increase in 1999 would be 8% of 19 956 = 1596.

4.1B

1 a 2.02%
b 53.33% of drivers involved in injury accidents were tested and 3.79% of these failed.

2 a $\frac{12\,970}{27\,122} \times 100\% = 48\%$ and $\frac{443}{12\,970} \times 100\% = 3\%$ to nearest per cent
b 1.63%

3 a i 0.452%
ii 1.14%
iii 0.495%
b This shows that there were far more car drivers than motorcycle riders involved in injury accidents (the combined percentage is weighted heavily towards the car driver percentage).

4 a i 93.1%
ii 6.92%
b i 94.3%
ii 5.70%
c Car drivers were more likely to fail/refuse a breath test than motorcycle riders that year.
Also more car drivers involved in injury accidents.

5 a i 3.70% **ii** 12.0%
b Greater increase in the % of motorcycle riders involved in injury accidents than car drivers.

6 a −8.68% (i.e. a decrease)
b 8.58%

4.1C

1 a 5123 thousand **b** 5120 thousand
c 5119 thousand **d** 5115 thousand
e 5111 thousand **f** 5107 thousand

2 a 49 089 thousand **b** 49 285 thousand
c 49 497 thousand **d** 49 754 thousand
e 49 998 thousand **f** 48 708 thousand
g 48 533 thousand **h** 48 379 thousand

3 a 1.010 44 to 3 s.f.
b 1.04 to 3 s.f.
c 0.863% to 3 s.f.

4 57 800 thousand
(% increase between 1991 and 2000 = 3.36% to 3 s.f.

4.1D

1 a £99 832.50 **b** £73 237.50
2 a £36 866.36 **b** £54 525.35
3 a 6.5% **b** 35.3% **c** 7.58%

4 England: prices drop before rising again after 1993; Wales: prices steady through early 90s before rising more sharply in late 90s; Scotland and Northern Ireland: steady increase through the decade.

4.2 Statistical charts and graphs

4.2A

1 a 1991
  b i about £620 ii about $900
  c (minimum in 1975, maximum in 1991)
    i about £600 ii about $865
  d 350%
  e All monetary values are given as 1975 equivalent amounts. This means that values are relative to 1975, not the absolute value in any particular year.

2 a i 1993 ii 1995
  b 75mm *per month*
  c 1984–1985

3 Drier summers coincide with increased claims in the same year quite noticeably in 1976, 1989–91, 1995–96, and affect the following year's claims as well (increasing effect).

4.2B

1 a 3.5–3.6 million
  b Many more females – roughly double the numbers of 80–84, and treble the number of 85+ males.
  c i 35–39 ii 1961–1965

2 a i Fewer
    ii Fewer
    iii Falling birth rate
    iv Increased death rate amongst the young.
  b i 66% approximately
    ii 50% approximately
    iii Females have a longer life expectancy.
  c i 85+
    ii Better general health through life, better health care and treatments available.

3 a i Apart from the actual sizes of each population group, the Indian population has a greater proportion of younger people whereas the UK population is more evenly distributed.
    ii UK is highly developed whereas India is developing rapidly and the birth rates associated with these types of development could account for the differences.
  b i Slightly more boys than girls in both countries.
    ii Very similar.

4 a Aging population, declining birth rate, improved health care will push people into the higher levels of the population pyramid, and reduce the numbers at the lower levels.
  b UK is shown with proportionally fewer children in its population than India, where the numbers are broadly even from ages 0–49. The UK also has a larger proportion of its population over 65 than India.

4.2C

1 a

	Price (pence)			
Day	Opening	Highest	Lowest	Closing
Wed	91	93	88	90
Thurs	90	108	89	105
Fri	107	112	95	101

  b −1.10%, +16.67%, −5.61%
  c i +10 pence
    ii +8 pence
  d i +10.99%
    ii +8.60%

2 Black Marubozu – price doesn't get any higher than when it opens and reaches its lowest price at closing.

Spinning Top – price varies greatly throughout the day's trading, but finishes up only a small amount (compared to the range of prices it was at throughout the day).

Long-legged Doji – price varies greatly throughout the day, finishing at practically the same value as its opening price.

Gravestone Doji – price increases during the day, and closes at practically the same price as it opened at.

3 a Price does not increase by much but falls greatly during the day but recovers to close at a little below the opening price.
  b Closes a little *above* the opening price.
  c Price falling over a few days then finally recovers a little on the final day to close above the opening price.

4 Share prices for both rose in the first week of June before falling to what would be the monthly low in the second week. Then MKS recovered better in the second half of June finishing close to where it started the month, whereas NXT price hovers around their low, finishing lower for the month.

Interpreting statistical diagrams

1 Increasing

Year	Consumption (billions of cans)	Recycling (millions of cans)	Recycling rate
1989	2.8	80	3%
1990	4.0	240	6%
1991	4.1	480	12%
1992	4.4	700	16%
1993	5.5	1160	21%
1994	6.4	1520	24%
1995	6.8	1900	28%
1996	4.9	1500	31%
1997	4.4	1520	35%
1998	4.3	1560	36%

2 Consumption grew steadily from 1989–95, but there has since been a steady fall in consumption. Recycling also grew steadily to a maximum number in 1995, but has stayed at a high level since.
3 Increased consumption linked to increased emissions.
4 Both are scatter diagrams where most points lie in a small region of the diagram with a small number of extreme points. Fig. 2 shows positive correlation i.e. as GDP increases there is an increase in CO_2 emissions, Fig. 3 shows very little correlation.
5 Any with correct interpretation of the figures to back up argument.
6 a i 118 cm ii 133 cm
b i 160 cm ii 10 cm
c 106% (from 86 to 177 cm)
d i 12–15 ii the gradient
7 a

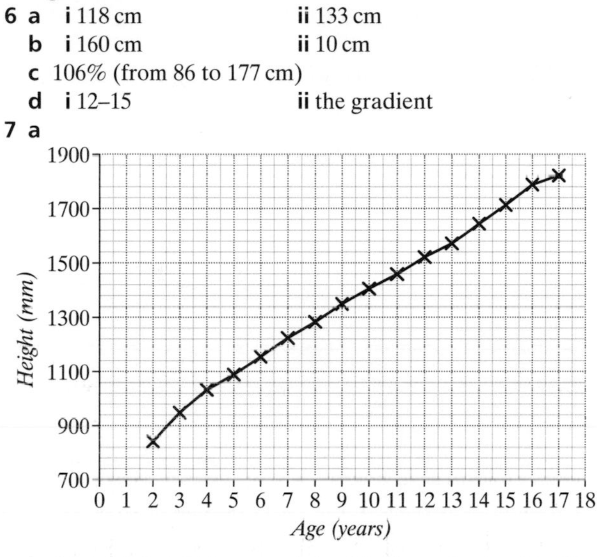

b Alistair's height increased steadily through ages 2 to 17, whereas in a 'typical' boy, there would be a more pronounced growth 'spurt' between 12 and 15 years.
8 a i approx 95p.
ii Prices shown are the closing prices – there will be a whole range of prices the shares were valued at during each day.
b 25%
c i 35 million ii £79 m (approx)
iii No trading at weekends.
d i Price has been falling from a high in mid-May, but has levelled out by November, to a little above half the price in mid-May.
ii Smooths out day to day fluctuations to provide a clearer trend.
iii There will be lag in any moving average, and decisions to buy and sell will be done from day to day at highs and lows and loss of detail of the original.
9 c ii 1.30 cm
d i 1350 deaths ii 1200 deaths

4.4 Preparing for assessment

Practice exam questions

1 a i 3.9 m^3s^{-1} ii 5 hours or 6 hours iii 6 hours
b i 4.7 mm ii 0.67 mm
c 975%
2 a i 4.1%
ii Similar proportions of pupils in the sample as the population in various groups, like boy/girl splits, ethnicity, family income.

b i 244 800 ii 252 000 iii 547 200
c Some of the students questioned took combinations of cigarettes, marijuana/cannabis and alcohol.
d i The populations of each of the 3 regions have to be taken into account to give a *weighted* average.
ii European students smoked cigarettes more and drank more alcohol than American students, but used marijuana/cannabis less.

5 Critical Thinking

This chapter is intended to generate discussion about the way others use statistics. Due to the subjective nature of the activities, set answers are not provided.

6 Extension opportunities

6.1 Stem and leaf diagrams

6.1B

1 a **Time spent on computers during a week by a group of students**

Males		Females
9 7	0	7 8 8 9
9 8 8 2 0	1	2 3 3 3 3 4 5 6 6 8
7 6 5 5 3 3 1 0	2	0 1 1 4 6 8
9 9 7 6 6 5 5 3 2 1 1	3	2 2 4 5
8 4 2	4	0 1 5
	5	0 6
3	6	3

Unit = 10 minutes

b Males are generally spending more time on computers than females – roughly double times the females spent.

2

Males		Females
7 2	8	4 5 7
9 9 8 8 7 6 6 5 3 3 3 2 0	9	3 4 4 4 5 6 7 8 8 8 9 9
9 8 6 5 4 4 4 2 0 0	10	0 0 1 2 3 3 4 5 7 7
5 4 3 3 0	11	0 3 4 6
	12	4

Unit = 1 IQ point

6.2 Box plots

6.2A

1 a

	Weight (kg)	
	Men	Women
Median	80	66
Min.	41	27
Max.	119	105
Range	78	78
LQ	71	57
UQ	89	76
IQR	18	19

2 Cars: over half over 70 mph
Cars towing: 25–50% over 60 mph
Motorcycles: half over 70 mph approx.
Light Goods: 25–50% over 70 mph
Coaches: <25% over 70 mph
HGVs: <25% over 60 mph

3 a

Statistic	Cars	Motorcycles	HGVs
LQ	66	66	52
Max.	102	98	62
Min.	54	48	41
UQ	78	81	58
Median	71	71	56

Any correct interpretation of these figures.

b Car speeds – reasonably symmetric except for a longer upper tail.

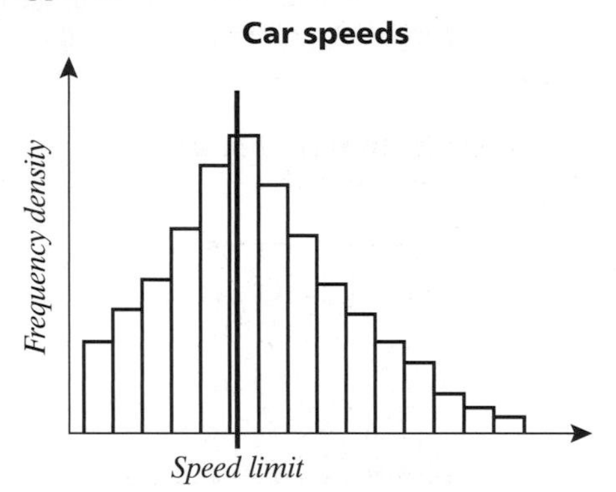

Motorcycle speeds – negatively skewed, with long upper tail.

Note that 25% of drivers are very close below the speed limit (LQ–median).

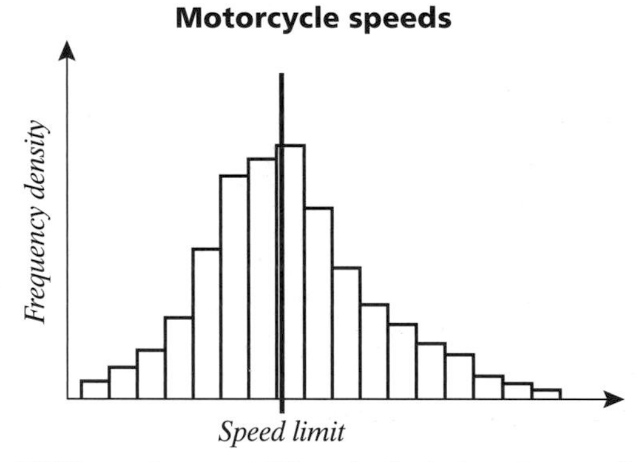

HGV speeds – over 75% under limit, long lower tail, sharp drop from median–UQ and short upper tail.

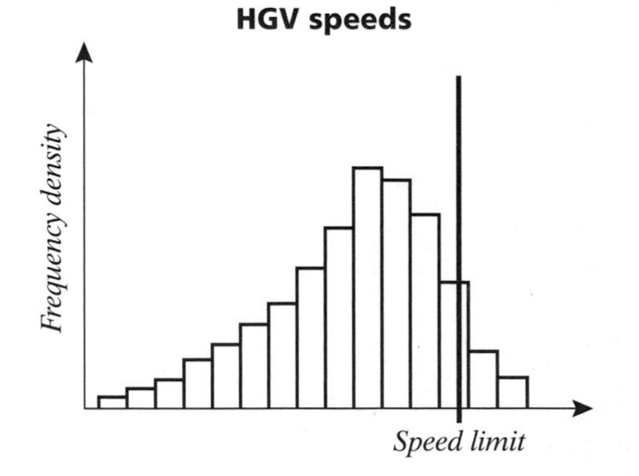

4 a Symmetric about median. Lengths the same from Max. to UQ to Median to LQ to Min.

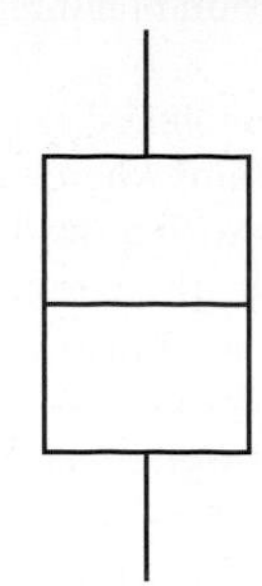

b LQ closer to Min. than UQ is from Max. Median closer to LQ than UQ.

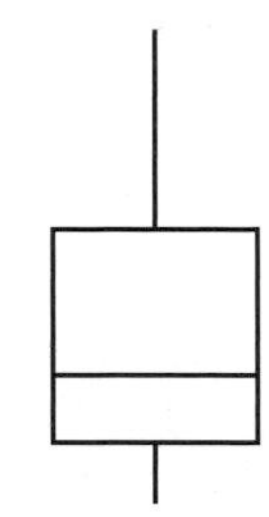

c Symmetric about median. Similar to uniform distribution, only the UQ and LQ are closer to the Max. and Min. respectively.

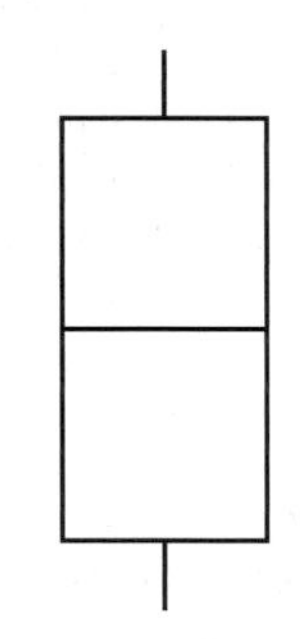

d UQ closer to Max. than LQ is from Min. Median closer to UQ than LQ.

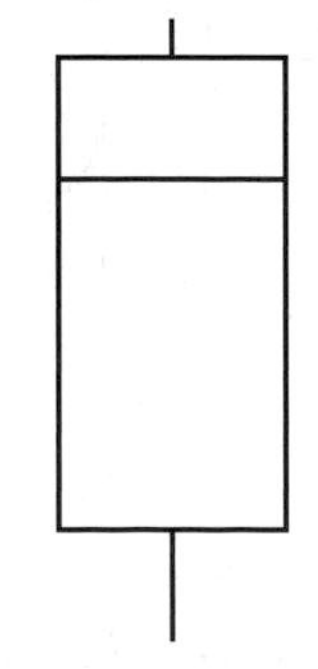

6.3 More about correlation and regression

6.3A

2 a 1

b -1

3 -0.547

4 a Weak

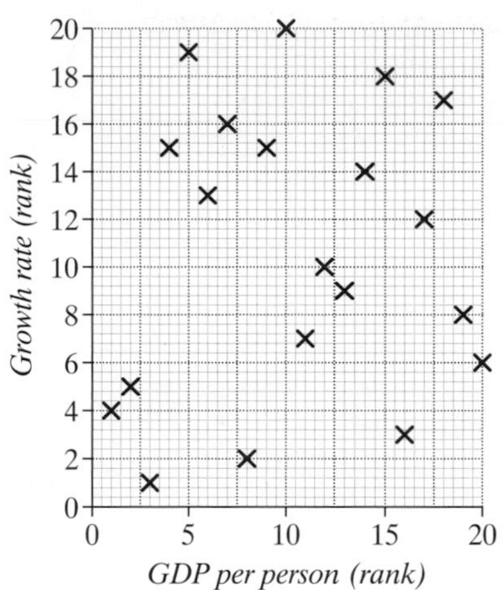

b 0.123

6.4 Probability distributions

6.4A

1 a All values 1 to 6 are equally likely on a single throw. The numbers of ways you can get the combined total scores from 2 to 12 vary. Only one way to get 2, whereas there are 6 ways to get a total of 7 so the probabilities are not equal.

b

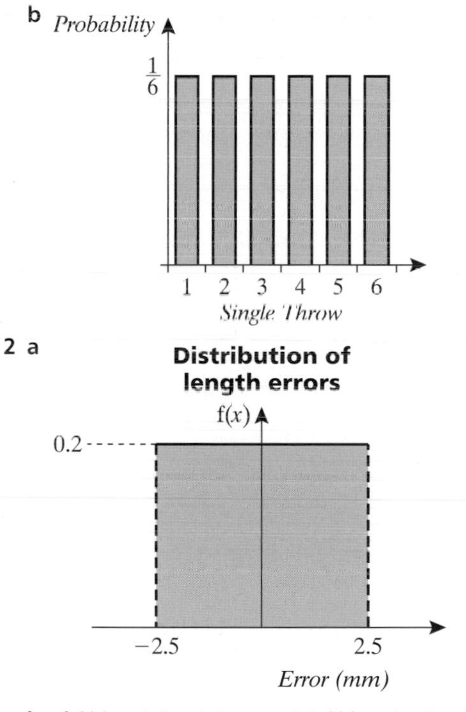

2 a

b **i** $f(x) = 0.2 \; -2.5 < x < 2.5 \; f(x) = 0$ otherwise
ii 0
iii 1.44 mm

3 a **i** Uniform from 0 to 180 degrees
ii Uniform from 0 to 1.2π (minor arc)

b

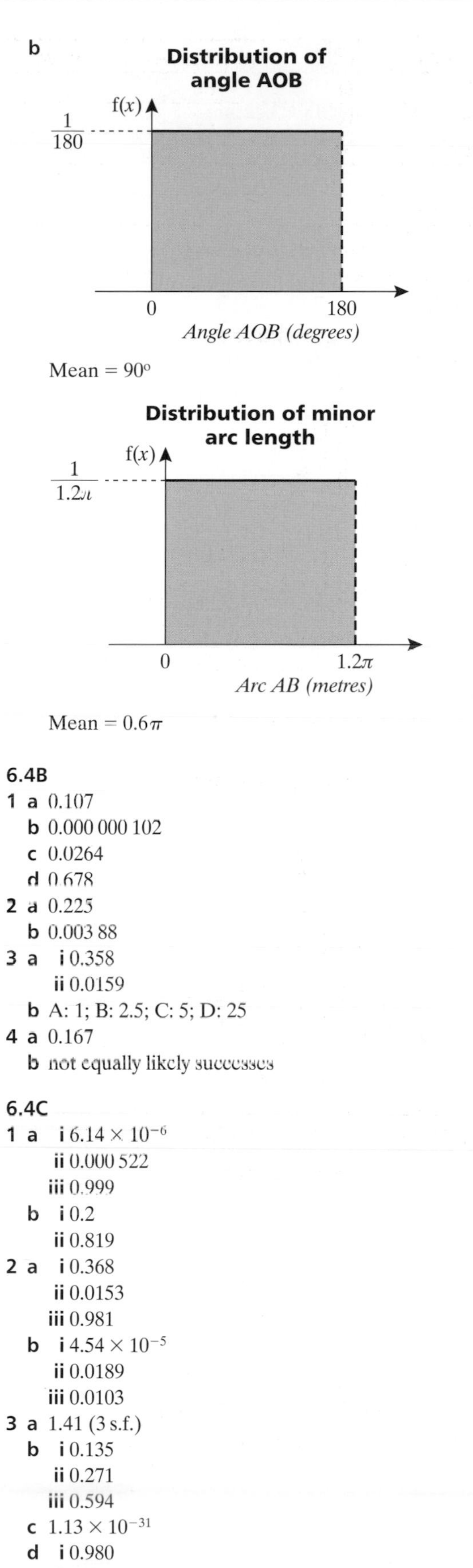

Mean = 90°

Mean = 0.6π

6.4B

1 a 0.107
b 0.000 000 102
c 0.0264
d 0.678

2 a 0.225
b 0.003 88

3 a **i** 0.358
ii 0.0159
b A: 1; B: 2.5; C: 5; D: 25

4 a 0.167
b not equally likely successes

6.4C

1 a **i** 6.14×10^{-6}
ii 0.000 522
iii 0.999
b **i** 0.2
ii 0.819

2 a **i** 0.368
ii 0.0153
iii 0.981
b **i** 4.54×10^{-5}
ii 0.0189
iii 0.0103

3 a 1.41 (3 s.f.)
b **i** 0.135
ii 0.271
iii 0.594
c 1.13×10^{-31}
d **i** 0.980
ii 0.0198

6.5 Significance tests

6.5A

1 b Mean = 1620 mm; SD = 3.22 mm

2 b Mean = 32 g; SD = 0.212 g

c Mean = 5000 hrs; SD = 14.2

6.5B

1 a Yes (test statistic = 2.235 > 1.65)

b No

c 95% confident that mean strength has increased, but not 99% certain

2 No (test statistic = 1.6 < 1.65)

3 a Yes (test statistic = –3.018 < –1.65)

b Yes (–3.018 < –2.33)

4 a No (test statistic = –8 < –2.33)

b Yes (test statistic = 12 > 2.33)

c Taking 501.8 as an estimate of
$U(f(X > 500) = P(Z > 1.2) = 88.5\%$
About 12% of packets likely to have less than the stated 500 g. Manufacturer not giving what they say to all customers.

6.5C

1 a Yes (test statistic = 2.07 > 1.96)

2 a Yes (test statistic = –2.37 < −1.96)

b No (critical value = –2.58 and −2.37 > −2.58)

c Change back to old display or try another way

Formulae

The product-moment correlation coefficient for bi-variate data:

$$r = \frac{S_{xy}}{\sqrt{S_{xx}S_{yy}}}, \text{ where } S_{xy} = \sum[(x_i - \bar{x})y_i]$$

$$S_{xx} = \sum[(x_i - \bar{x})x_i]$$

$$S_{yy} = \sum[(y_i - \bar{y})y_i]$$

Spearman's rank correlation coefficient: $r = 1 - \dfrac{6\Sigma d^2}{n(n^2 - 1)}$

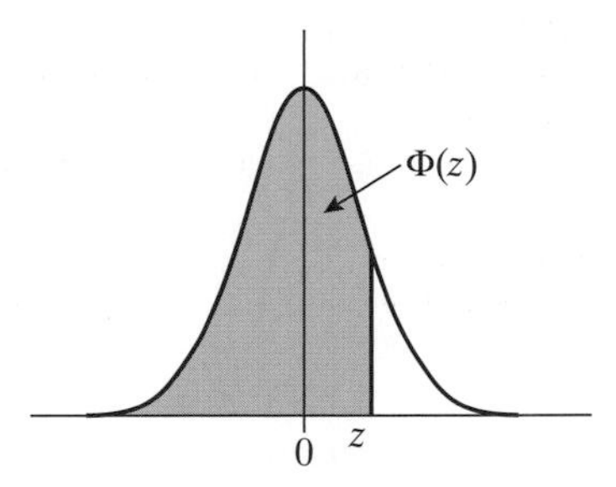

Normal Distribution

The tabulated value is $\Phi(z) = P(Z \leq z)$ where Z is the standardised normal random variable, $N(0, 1)$.

z	0	0.01	0.02	0.03	0.04	0.05	0.06	0.07	0.08	0.09
0.0	0.5000	0.5040	0.5080	0.5120	0.5160	0.5199	0.5239	0.5279	0.5319	0.5359
0.1	0.5398	0.5438	0.5478	0.5517	0.5557	0.5596	0.5636	0.5675	0.5714	0.5753
0.2	0.5793	0.5793	0.5793	0.5793	0.5793	0.5793	0.5793	0.5793	0.5793	0.5793
0.3	0.6179	0.6217	0.6255	0.6293	0.6331	0.6368	0.6406	0.6443	0.6480	0.6517
0.4	0.6554	0.6591	0.6628	0.6664	0.6700	0.6736	0.6772	0.6808	0.6844	0.6879
0.5	0.6915	0.6950	0.6985	0.7019	0.7054	0.7088	0.7123	0.7157	0.7190	0.7224
0.6	0.7257	0.7291	0.7324	0.7357	0.7389	0.7422	0.7454	0.7486	0.7517	0.7549
0.7	0.7580	0.7611	0.7642	0.7673	0.7704	0.7734	0.7764	0.7794	0.7823	0.7852
0.8	0.7881	0.7910	0.7939	0.7967	0.7995	0.8023	0.8051	0.8078	0.8106	0.8133
0.9	0.8159	0.8186	0.8212	0.8238	0.8264	0.8289	0.8315	0.8340	0.8365	0.8389
1.0	0.8413	0.8438	0.8461	0.8485	0.8508	0.8531	0.8554	0.8577	0.8599	0.8621
1.1	0.8643	9.8665	0.8686	0.8708	0.8729	0.8749	0.8770	0.8790	0.8810	0.8830
1.2	0.8849	0.8869	0.8888	0.8907	0.8925	0.8944	0.8962	0.8980	0.8997	0.9015
1.3	0.9032	0.9049	0.9066	0.9082	0.9099	0.9115	0.9131	0.9147	0.9162	0.9177
1.4	0.9192	0.9207	0.9222	0.9236	0.9251	0.9265	0.9279	0.9292	0.9306	0.9319
1.5	0.9332	0.9345	0.9357	0.9370	0.9382	0.9394	0.9406	0.9418	0.9429	0.9441
1.6	0.9452	0.9463	0.9474	0.9484	0.9495	0.9505	0.9515	0.9525	0.9535	0.9545
1.7	0.9554	0.9564	0.9573	0.9582	0.9591	0.9599	0.9608	0.9616	0.9625	0.9633
1.8	0.9641	0.9649	0.9656	0.9664	0.9671	0.9678	0.9686	0.9693	0.9699	0.9706
1.9	0.9713	0.9719	0.9726	0.9732	0.9738	0.9744	0.9750	0.9756	0.9761	0.9767
2.0	0.9772	0.9778	0.9783	0.9788	0.9793	0.9798	0.9803	0.9808	0.9812	0.9817
2.1	0.9821	0.9826	0.9830	0.9834	0.9838	0.9842	0.9846	0.9850	0.9854	0.9857
2.2	0.9861	0.9864	0.9868	0.9871	0.9875	0.9878	0.9881	0.9884	0.9887	0.9890
2.3	0.9893	0.9896	0.9898	0.9901	0.9904	0.9906	0.9909	0.9911	0.9913	0.9916
2.4	0.9918	0.9920	0.9922	0.9925	0.9927	0.9929	0.9931	0.9932	0.9934	0.9936
2.5	0.9938	0.9940	0.9941	0.9943	0.9945	0.9946	0.9948	0.9949	0.9951	0.9952
2.6	0.9953	0.9955	0.9956	0.9957	0.9959	0.9960	0.9961	0.9962	0.9963	0.9964
2.7	0.9965	0.9966	0.9967	0.9968	0.9969	0.9970	0.9971	0.9972	0.9973	0.9974
2.8	0.9974	0.9975	0.9976	0.9977	0.9977	0.9978	0.9979	0.9979	0.9980	0.09981
2.9	0.9981	0.9982	0.9982	0.9983	0.9984	0.9984	0.9985	0.9985	0.9986	0.9986
3.0	0.9987	0.9987	0.9987	0.9988	0.9988	0.9989	0.9989	0.9989	0.9990	0.9990
3.1	0.9990	0.9991	0.9991	0.9991	0.9992	0.9992	0.9992	0.9992	0.9993	0.9993
3.2	0.9993	0.9993	0.9994	0.9994	0.9994	0.9994	0.9994	0.9995	0.9995	0.9995
3.3	0.9995	0.9995	0.9995	0.9996	0.9996	0.9996	0.9996	0.9996	0.9996	0.9997
3.4	0.9997	0.9997	0.9997	0.9997	0.9997	0.9997	0.9997	0.9997	0.9997	0.9998
3.5	0.9998	0.9998	0.9998	0.9998	0.9998	0.9998	0.9998	0.9998	0.9998	0.9998
3.6	0.9998	0.9998	0.9999	0.9999	0.9999	0.9999	0.9999	0.9999	0.9999	0.9999
3.7	0.9999	0.9999	0.9999	0.9999	0.9999	0.9999	0.9999	0.9999	0.9999	0.9999

$$f(x) = \frac{1}{\sigma\sqrt{2\pi}} \exp\left(-\frac{(x-\mu)^2}{2\sigma^2}\right)$$

Index

BIG MAC ECONOMICS

How can you compare the economies of different countries around the world?

Well, The Economist magazine suggests that the cost of a Big Mac from MacDonald's can help. Every year, in April, The Economist updates its Big Mac Index. The April 2002 results for some countries are shown in the spreadsheet in Figure 2.

Figure 1: A Big Mac

	A	B	C	D	E	F	G
1				Big Mac price			
2	Country	Local currency	Exchange rate (local currency per $)	In local currency	In dollars	PPP	under or over valuation %
3	USA	Dollar ($)	1	2.49	2.49	1.00	0%
4	Britain	Pounds (£)	0.69	1.99	2.89	0.80	16%
5	Canada	C Dollar	1.57	3.33	2.12	1.34	-15%
6	China	Yuan	8.28	10.50	1.27	4.22	-49%
7	Hong Kong	HK Dollar	7.80	11.20	1.44	4.50	-42%
8	Indonesia	Rupiah	9430	16000	1.70	6426	-32%
9	Japan	Yen	130	262	2.02	105	-19%
10	Mexico	Peso	9.28	21.90	2.36	8.80	-5%
11	New Zealand	NZ dollars	2.24	3.95	1.76	1.59	-29%
12	Russia	Rouble	31.20	39.00	1.25	15.66	-50%
13	South Africa	Rand	10.90	9.70	0.89	3.90	-64%
14	Switzerland	Swiss Franc	1.66	6.30	3.80	2.53	52%

Figure 2: Big Mac Index (April 2002)

This gives a lot of information for each country. First of all column C gives the exchange rate – ie, the amount of local currency you would have got in April 2002 for one US Dollar.

For example, in Britain at that time you would have received £0.69 or 69 pence for one dollar.

The price of a Big Mac in each country is then given in the local currency (column D) – in Britain at that time a Big Mac cost £1.99. This is converted to a price in US Dollars, in column E, by dividing by the exchange rate.

Figure 3: MacDonald's restaurant

For example, in Britain:

$$\frac{\text{Big Mac price in local currency}}{\text{Exchange rate}} = \frac{1.99}{0.69} = 2.89$$

This tells you that a Big Mac in Britain effectively costs $2.89, (2.89 US Dollars). This means that a visitor to Britain from the USA will be paying more in dollars for a Big Mac in Britain than he or she would at home.

For each country column F gives the Big Mac Index which is based on the economic theory of purchasing-power parity (PPP). In this case when considering the Big Mac this means that you assume that a Big Mac is worth the same (has parity) in each country and therefore should cost the same in each country. The PPP or Big Mac Index is found by dividing the Big Mac price in the country's local currency by the cost of a Big Mac in the US.

For example in Japan, a Big Mac cost 262 Yen. Dividing this by the cost of a Big Mac in the US (i.e. 2.49 dollars) gives 105. This means that if a Big Mac has the same value in Japan as in the USA then 1 US dollar should be worth 105 Yen. However, as the table shows 1 US dollar was exchanged for 130 Yen at the time, which means that effectively an American Tourist receives 25 Yen too many for each dollar. The Japanese Yen is therefore under-valued by

$$\frac{(130 - 105)}{130} \times 100\% = 19\%$$

The final column (G), therefore, gives the percentage by which the local currency is under- or over-valued using the Big Mac as the unit of currency.

This Big Mac Index is a useful indicator of how world economies are performing. Economists argue that in the long run exchange rates and the price of Big Macs in different countries will move so that you would effectively pay the same for a Big Mac in whichever country you happen to be.

The box plot in Figure 4 is based on the price (in US dollars) of a Big Mac in 31 countries around the world in 2002. It shows the median – the line in the box – as a measure of location and the minimum and maximum values and lower and upper quartiles as measures of spread. The lower and upper quartiles define the edges of the box and the minimum and maximum values are at the ends of the whiskers (the lines that extend from the box).

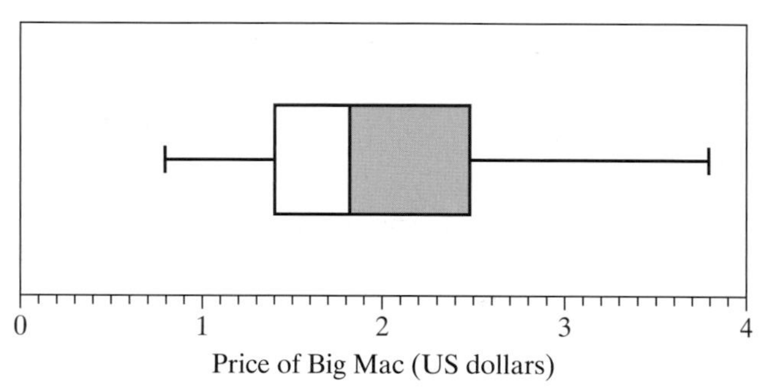

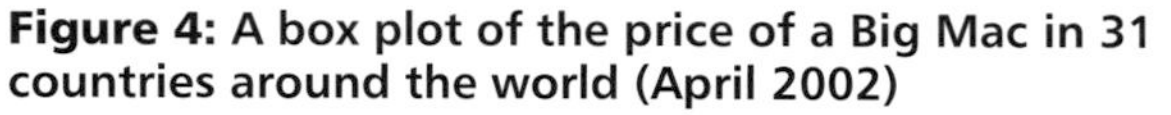

Figure 4: **A box plot of the price of a Big Mac in 31 countries around the world (April 2002)**

The table in Figure 5 shows how the Big Mac Index for a number of countries varied in the period 1989–2002. Trends in these data are seen more easily by looking at line graphs that show how the

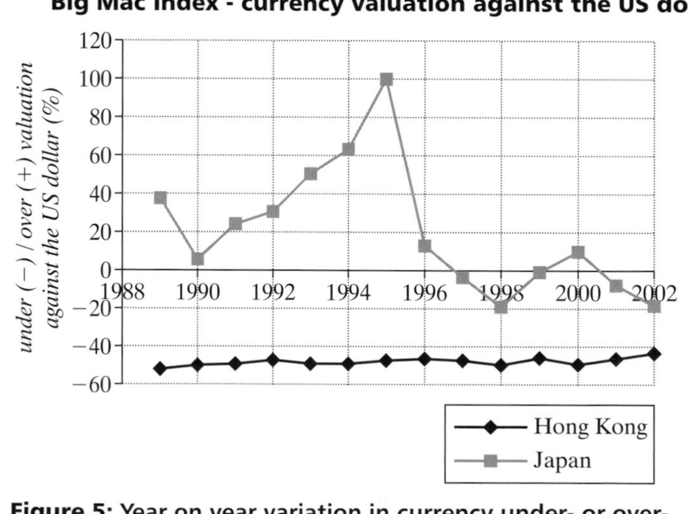

Figure 5: **Year on year variation in currency under- or over-valuation against the US dollar.**

Big Mac Index for a country varies year by year. Such graphs for four countries are shown in Figure 6. These show that, for example:

- the British Pound has been consistently over-valued against the US dollar
- the Japanese Yen has fluctuated a great deal in value when compared to the US dollar.

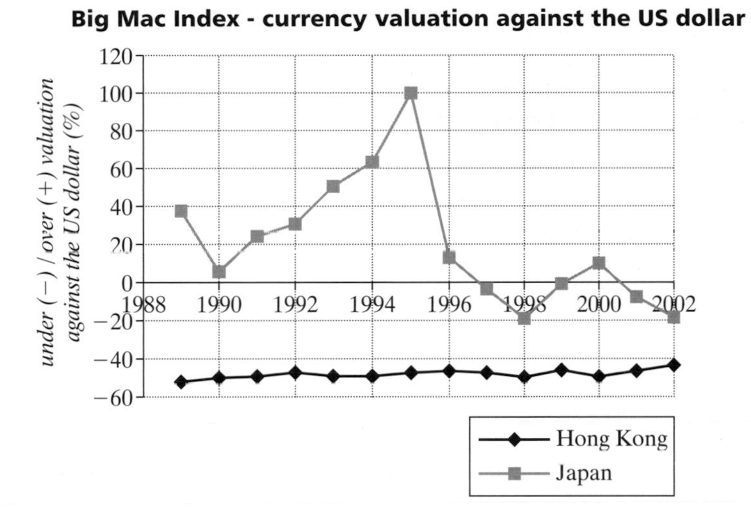

Figure 6: Graphs showing yearly currency under- or over-valuation against the US dollar based on the Big Mac Index.

Comprehension Questions

1. In April 2002 a Big Mac in Australia cost 3.00 Australian dollars. At that time 1 US dollar was worth 1.86 Australian dollars.
   For April 2002, calculate
   a. the cost of an Australian Big Mac in US dollars;
   b. the Big Mac Index for Australia;
   c. the amount by which the Australian dollar was under- or over-valued when compared with the US dollar.
2. If the British pound had been neither under- nor over-valued against the US dollar in April 2002 how much would a Big Mac have cost in Britain?
3. a. Use the box plot in Figure 4 to find estimates of:
      i the median,
      ii the range,
      iii the interquartile range
      of the prices of Big Macs around the world.
   b. Comment on what the box plot shows about the prices around the world, referring to measures of location and spread in your answer.
4. A travel magazine states that American tourists are well off when visiting overseas countries.
   a. Explain how Figure 4 supports this conclusion.
   b. Use Figure 4 to estimate the proportion of countries in which American Tourists will pay more for a Big Mac than in the US.
5. Use Figure 6 to identify a country whose currency is tightly tied to the strength of the US dollar. Explain your choice.
6. Use data from Figure 5 to draw a graph showing the under- or over-valuation of Canadian dollar against the US dollar and comment on what the graph shows.
7. Use Figure 5 to identify when the communist government of Russia collapsed. Explain your choice.

IS YOUR DATA NORMAL

When you have data from a sample of a population how can you tell if it is normal? For example, the table in Figure 1 shows the heights of nine 15-year-old girls. We might expect that the distribution of girls' heights at a particular age can be modelled by a normal probability model. But how do you check?

height, h cm
159
160
160
162
162
167
168
170
170

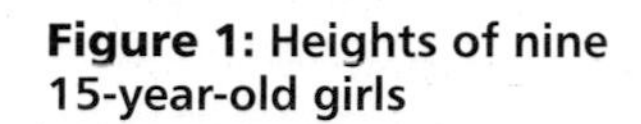

Figure 1: **Heights of nine 15-year-old girls**

One method you can use is to first of all put the data in ascending order and to investigate whether these values plotted against the corresponding standard normal quantiles lie close to a straight line. It is perhaps easier to understand this method by following it for the example of the girls' height data.

In this case there are nine data values and if these were ideally normally distributed you would expect that the first height (when the heights are arranged in ascending order) would correspond to the value, z_1, given by $\Phi(z_1) = \frac{1}{10}$ (see Figure 2). You would expect the second height to correspond to the value z_2, given by $\Phi(z_2) = \frac{2}{10}$ and so on.

The values $z_1, z_2, \ldots z_{10}$ are the standard normal deciles in this case.

If you have n data values you plot these against the corresponding standard normal quantiles, z_i, given by solving the equation $\Phi(z_i) = \frac{i}{n+1}$. So if you have only three data values you would plot these against the corresponding normal quartiles.

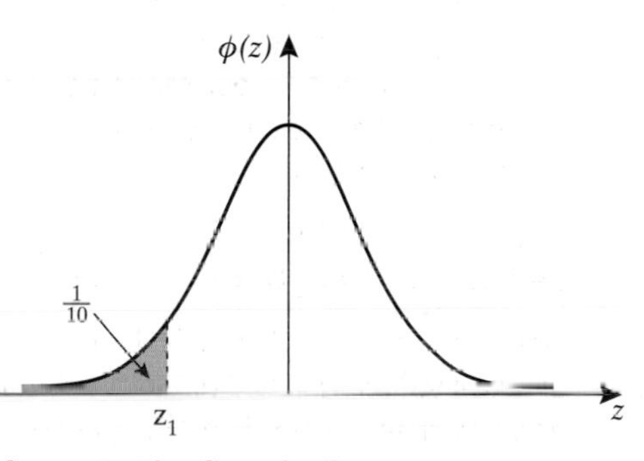

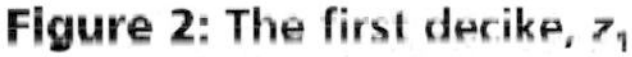

Figure 2: **The first decike, z_1**

The table in Figure 3 gives the data values for the girls' heights and the corresponding standard normal deciles.These values are shown plotted in the graph of Figure 4.

i	h_i	$\frac{i}{10}$	x_i
1	159	0.1	−1.282
2	160	0.2	−0.842
3	160	0.3	−0.524
4	162	0.4	−0.253
5	162	0.5	0.000
6	167	0.6	0.253
7	168	0.7	0.524
8	170	0.8	0.842
9	170	0.9	1.282

Figure 3: **Nine data values and their corresponding deciles, x_i**

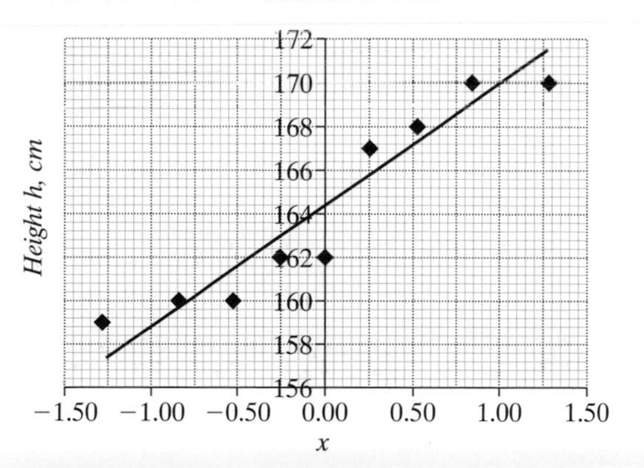

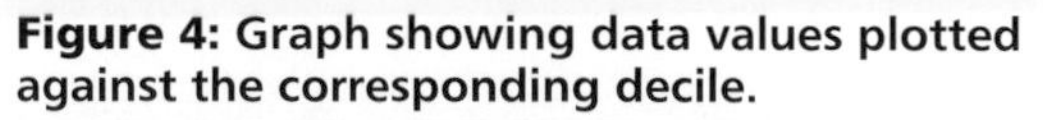

Figure 4: **Graph showing data values plotted against the corresponding decile.**

You can see that in this case the heights plotted against the standard normal deciles lie quite close to a straight line. Using the statistical functions of your calculator you can find the equation of the regression line to be $h = 5.198x + 164.2$ and the correlation coefficient to be 0.949. This value of correlation coefficient certainly confirms that there is a high degree of correlation between the girls' height data and the standard normal deciles. The equation of the straight line ($h = 5.198x + 164.2$) gives the intercept on the h-axis as 164.2, which is an estimate of the value of the mean, median and modal height of the population. The gradient of the straight line (5.198) gives an estimate of the standard deviation, since the values on the x-axis are effectively standard deviations from the mean. Therefore in this particular case the data can be modelled as being from a population with mean height 164.2 cm and standard deviation 5.198 cm.

This method is useful for situations where you have a relatively small sample of data from a population and want to find whether the data fits an underlying normal model. On other occasions you may have large amounts of data and still need to decide whether it is 'normal'. The table in Figure 5 shows the cumulative percentage of adult females having various eye-heights when seated (see Figure 6).

This type of data is particularly useful to furniture designers – for example, when designing a computer workstation. The data is shown graphed in Figure 7, giving a typical cumulative percentage graph.

Sitting Eye Height, mm	Cumulative Percentages
1000	0
1020	0.8
1040	3.2
1060	6.2
1080	11.8
1100	21.8
1120	35.6
1140	53.4
1160	69.4
1180	82.2
1200	89.4
1220	94.8
1240	98
1260	99.4
1280	100

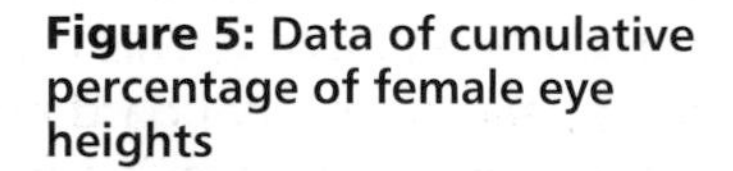

Figure 5: Data of cumulative percentage of female eye heights

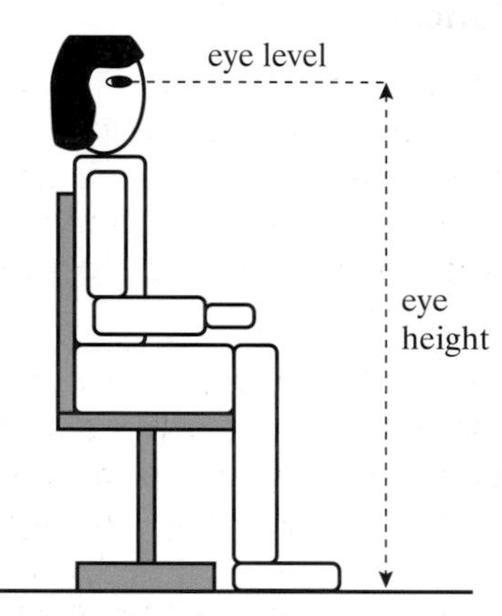

Figure 6: Female eye height defined

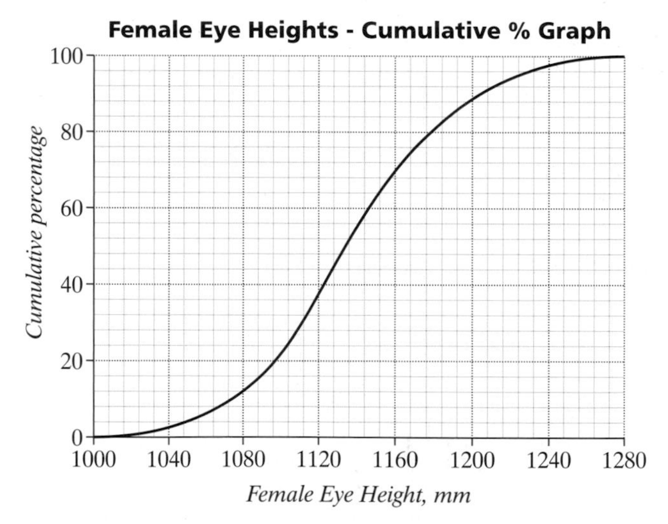

Figure 7: Cumulative percentage plot of female eye heights

If this data is normally distributed, is this the type of graph you would expect? To decide you need to plot $\Phi(z)$ against z (see Figure 8). This is shown in Figure 9 below.

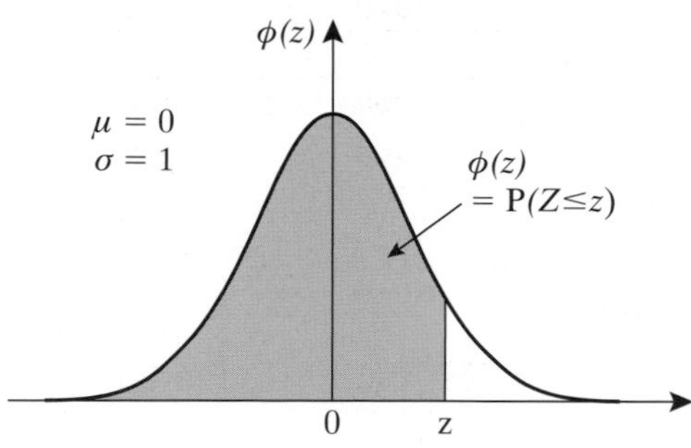

Figure 8: The standard normal distribution

Standard Normal Cumulative Distribution Function

$\phi(z)$

Figure 9: Graph of $\Phi(z)$ plotted against z

The shaded area in the sketch represents the probability that a random value taken from the standard normal distribution is less than or equal to z. This is the value that appears in the standard normal table as . is called the **cumulative distribution function**.

For values of z below -3, $\Phi(z)$ is very small. When z reaches 3 the value of $\Phi(z)$ is very near 1. The gradient of the curve increases as z increases from -3 to 0 and then decreases again. The graph is steepest near the centre because this is where most of the values of the distribution lie.

From this you can deduce that any data that produces a 'standard' cumulative frequency graph may be modelled by a normal distribution. However you can check more quickly and easily by plotting cumulative percentages of the data on normal probability paper. The vertical scale on such paper has been devised so that a sample from a normal distribution gives a straight line. The graph in Figure 10 shows what this scale looks like. The scale is based on the inverse of the standard normal cumulative distribution function.

Points have been plotted on the graph showing the cumulative percentages from the sample of female eye heights. Note how close the points lie to a straight line. This suggests that the normal distribution provides a very good model for these particular data.

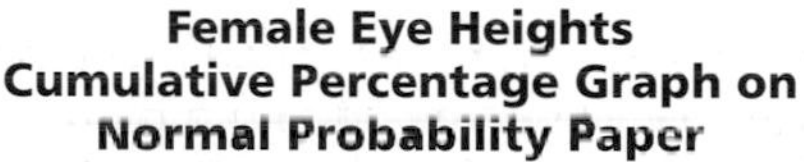

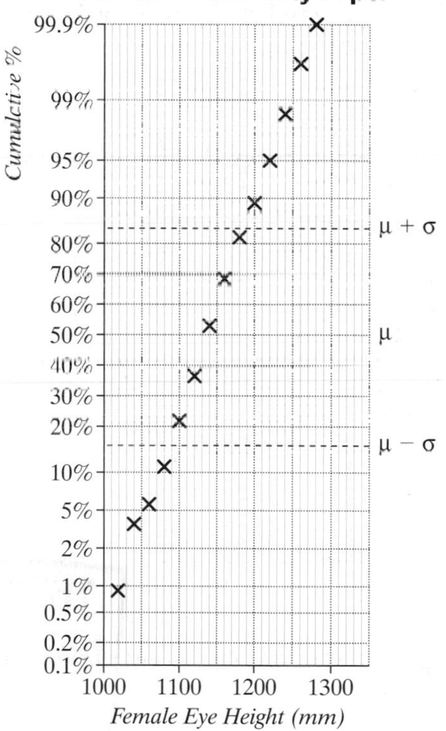

Figure 10: Female eye height plotted on normal probability paper

Comprehension Questions

1 a Show clearly the calculations required to confirm that $x_9 = 1.282$ when $\Phi(x_9) = 0.9$.
  b Explain why $\Phi(-1.282) = 0.1$ if $\Phi(-1.282) = 0.9$.

2 Show clearly why the statement,

"If you have n data values you plot these against the corresponding standard normal quantiles, x_i, given by

solving the equation $\Phi(x_i) = \frac{i}{n+1}$,"

leads to the sentence which follows, that is,

"So if you have only three data values you would plot these against the corresponding normal quartiles."

3 Use the equation $h = 5.198x + 164.2$ to find the maximum and minimum heights of 15 year-old girls who lie within one standard deviation of the mean height if their heights can be modelled by the normal distribution suggested in the article.

4 Explain why the gradient of the straight line $h = 5.198x + 164.2$ gives an estimate of the standard deviation.

5 From the graph in Figure 7 find
  a the median eye-height
  b the interquartile range of eye heights of adult females

6 Use tables to find the value of z for which a standard normal probability distribution has a cumulative probability of 0.8.

7 The article states that the graph of the standard normal distribution function "is steepest near the centre because this is where most of the values of the distribution lie."
Explain this statement.

8 a In the diagram, a line of best fit has been added to the cumulative percentage graph plotted on normal probability paper of Figure 10. Use it to find values for
    i μ, and
    ii σ.
  b Use the graph to write down estimates of:
    i the lower quartile, and
    ii the upper quartile.
  Explain how you found these.

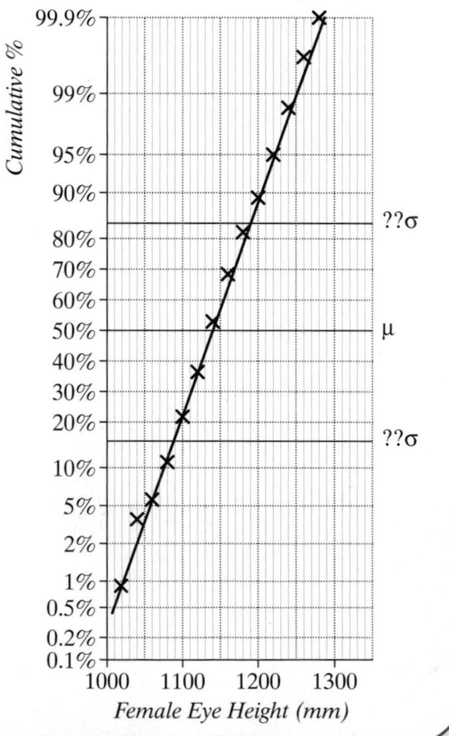